Possible Dreams
Enthusiasm for Technology in America

The Contributors

Joseph J. Corn teaches courses on American culture and the history of technology at Stanford University. His publications include *The Winged Gospel: America's Romance with Aviation, 1900–1950,* and *Yesterday's Tomorrows: Past Visions of the American Future.*

Susan J. Douglas is Associate Professor of Media and American Studies at Hampshire College. She is the author of *Inventing American Broadcasting, 1899–1922,* and serves on the Executive Council of the Society for the History of Technology.

Robert C. Post is a curator at the National Museum of American History and Editor-in-Chief of *Technology and Culture,* the quarterly journal of the Society for the History of Technology. He has written or edited nine books and is currently at work on an interpretation of drag racing, to be published by The Johns Hopkins University Press.

Carroll Pursell is Adeline Barry Davee Professor of History at Case Western Reserve University in Cleveland, Ohio. He was elected to serve as President of the Society for the History of Technology for 1990–92, and in 1991 was awarded the Society's Leonardo da Vinci Medal for lifetime contributions to the field.

Mary L. Seelhorst served as Curator and Coordinator of the Henry Ford Museum & Greenfield Village exhibition "Possible Dreams: *Popular Mechanics* and America's Enthusiasm for Technology." She received her master's degree in folklore from the University of North Carolina at Chapel Hill.

Michael L. Smith teaches the history of technology and 20th-century U.S. history at the University of California at Davis. His publications include *Pacific Visions: California Scientists and the Environment, 1850–1915,* as well as numerous articles. He is currently at work on a book titled "Special Effects," which traces representations of technology in Cold War America.

John L. Wright is Director of Educational Advancement at Henry Ford Museum & Greenfield Village. He has been involved in a variety of exhibition and publication projects at the museum, including "Streamlining America" and "The Automobile in American Life." He was a member of The Michigan Council for the Humanities from 1986 to 1990, and served as its chair from 1987 to 1989.

POSSIBLE DREAMS

Enthusiasm for Technology in America

Edited and with an Introduction by
John L. Wright

Henry Ford Museum & Greenfield Village
Dearborn, Michigan

Possible Dreams: Enthusiasm for Technology in America

Edited by Fannia Weingartner
Designed by Sharon Blagdon-Smart
Photographic Research by John Tobin
Photography by Rudy T. Ruzicska, Alan R. Harvey, and Tim Hunter
Composition by Paragraphics, Inc.
Printed by Edwards & Broughton Co.

Library of Congress Catalog Card Number: 91-077033
ISBN: 0-933728-35-2
© 1992 by Henry Ford Museum & Greenfield Village
All rights reserved
Printed in the United States

Front cover: (clockwise) *Popular Mechanics* cover, April 1932; Ford Model T, 1911; Prismatic disc transmitter invented by C. Francis Jenkins in 1923; Motorola microprocessor 60820, introduced 1984.

Contents

HENRY FORD MUSEUM & GREENFIELD VILLAGE

Henry Ford Museum & Greenfield Village is a national museum of American history and technology which has as its mission to collect, preserve, and interpret to a broad public audience the American historical experience, with a special emphasis on the relationship between technological change and American history.

The goal of all Henry Ford Museum & Greenfield Village programs is to challenge diverse audiences to reach beyond the limitations of their personal experience to a richer understanding of human possibilities, illuminated by historical understanding.

All the institution's programs derive from its extraordinarily rich collections comprising more than one million artifacts and twenty-five million books, manuscripts, prints and photographs, as well as other paper materials. These provide an unparalleled documentation of the history of American transportation, agriculture, industry, communication, domestic life, and leisure and entertainment. Taken together, the collections of Henry Ford Museum & Greenfield Village are the best of their kind in the world.

Henry Ford Museum & Greenfield Village is an independent, non-profit educational institution, not affiliated with the Ford Motor Company or the Ford Foundation.

Preface

This book, published on the occasion of a special exhibition on the same subject presented at Henry Ford Museum, explores a variety of topics relating to the history of enthusiasm for technology in America.

This general topic is an especially appropriate one for Henry Ford Museum & Greenfield Village. The focus of the Museum collections and educational program is on change and innovation, and on the achievements of those who helped bring them about. Thomas Edison, Orville and Wilbur Wright, George Washington Carver, Luther Burbank, Charles Steinmetz, and Henry Ford himself were all proponents of the benefits of technology. Each saw its effects as liberating rather than as enslaving. Their attitude scarcely differed from that of millions of other Americans, whose enthusiasm for technology was a basic tenet of their faith in democracy. They believed that the right tools and the right system could solve almost any problem.

While there is much historical evidence and a continuing current of critical literature that challenge the claims of enthusiastic believers in the unlimited benefits of technology, that belief cuts deeply through the American experience. We hope that the essays in this volume will help illuminate some facets of this important and traditional "American" view of the relationship between technology and the good life.

Harold K. Skramstad, Jr.
President

Acknowledgments

A book of this kind, like the major exhibition it parallels, is a complex enterprise requiring the contribution and collaboration of a great many people. All deserve to be thanked.

Within the Museum, Harold K. Skramstad, Jr., President, must be thanked for his support and encouragement of the book and exhibit projects. Steven K. Hamp, Director of Program, was the person who first saw the potential of these projects and played the key role in making them happen.

The exhibit team's conceptual planning and research have greatly influenced the book. For the exhibit, Donna R. Braden served as a Curator and Project Manager, and Mary L. Seelhorst as Project Curator and Coordinator. Important additional curatorial roles were filled by Peter H. Cousins, Randy R. Mason, and John Wesley Hardin. We especially appreciate their help in choosing and researching a number of the artifacts pictured in the book.

We are grateful to *Popular Mechanics* magazine for support of the exhibit project and for permission to use illustrations from the magazine in this book.

I am indebted to William S. Pretzer, Curator and Chairman of the Museum Publications Committee, for his help in identifying an outstanding group of authors and for his thoughtful critique of my Introduction.

John Tobin has done a fine job in researching historical photographs and in coordinating many of the details of production of the publication. Delores E. Hilbush has also been crucial in assisting with the logistics of book preparation.

The Museum's Photographic Department, headed by Rudy T. Ruzicska, is responsible for many of the excellent graphic reproductions. Staff include Alan Harvey, Tim Hunter, Virginia Morrow, Lisa Bachman Vitale, and Krista Boote.

Fannia Weingartner, Publications Consultant to the Museum, has as much right as I do to have her name on the cover as the book's editor. She has been the prime mover and shaper of the publication, from conceptualization through editing and on to production management. As designer, Sharon Blagdon-Smart has again provided us with a visually attractive and readable book.

Finally, I wish to thank my wife Sandra and my son Zach for their forbearance with my work on this project, including writing while on our family vacation. My thanks go as well to my mother Alice and my father John B. Wright, who have always shown me by their example that the American spirit of self-reliance and intelligent ingenuity is alive and well.

John L. Wright
Director of Educational Advancement

Introduction: An American Tradition

JOHN L. WRIGHT

This book and the exhibit it parallels have been occasioned in part by the 90th anniversary in 1992 of *Popular Mechanics* magazine. They have also been prompted by a growing body of writing on the history of American attitudes toward technology. One of the most pertinent and comprehensive of these works, published in 1989, is Thomas P. Hughes's *American Genesis*, subtitled "A Century of Invention and Technological Enthusiasm, 1870–1970." Hughes defines technology as "the effort to organize the world for problem solving so that goods and services can be invented, developed, produced, and used." He identifies the values of order, system, and control—embedded in machines, devices, and processes—as the key characteristics of enthusiasm for technology. Hughes applies this broad definition of technology particularly to the development of large modern complex systems such as automobile production and electric power generation and distribution.

However, we can see the workings of this enthusiasm from the very beginnings of white settlement in America. Captain John Smith's lurid story of his capture by Indians and his salvation from death by Pocahontas was first published in his *Generall Historie of Virginia* in 1624. His book has long been recognized as filled with sentimental and sensational exaggeration, yet its stories became part of the American mythic landscape.

One of the recurring themes in Smith's account is his ability to forestall the wrath of the Native Americans by technological wizardry. During his captivity he diverted them by giving their "King" his "round Ivory double compass Dyall." He noted

Facing page: The giant steam engine built by George Corliss for the 1876 Centennial Exposition in Philadelphia was a central symbol of America's enthusiasm for technology in the late 19th century.

that "Much they marvailed at the playing of the Fly and Needle" and that at his explanation of the scientific nature of the universe "they all stood as amazed with admiration."

Later in his history, Smith recounted the effect he produced on the Indians, who had in the meantime become friendly and were carrying millstones away from the English camp, by firing off several cannons:

> But when they did see him discharge them, being loaded with stones, among the boughs of a great tree loaded with Isickles, the yce and branches came so tumbling downe, that the poore Salvages [sic] ran away halfe dead with feare.

Smith delighted in the cultural hegemony established by his technically advanced firepower. It is a situation that would be repeated many times over the next two centuries as Europeans conquered Native Americans by means of superior technology in weaponry.

As Americans continued to settle and develop the continent, they came to view themselves and their country as having a character distinct from their European origins. The historian Frederick Jackson Turner in his influential essay of 1893, "The Significance of the Frontier in American History," asserted that the frontier experience had been the defining factor in shaping the American character. While nowadays Turner's thesis is considered too restrictive to serve as a comprehensive explanation, it continues to provide a useful framework for understanding the way in which members of the dominant cultural group of West European origin defined themselves.

The foundation of this value system was individualism, the belief that each person's freedom, needs, and rights should generally take precedence over the needs of the state or the larger society. The

The early English colonists at Jamestown intimidated the Native Americans by firing off their "great gunns."

continuing American frontier experience had shaped this belief through the practical need for pioneers to be self-reliant, independent, and resourceful in order to survive and prosper. The need for resourcefulness in living in the new land had fostered a spirit known as "Yankee ingenuity," which gave Americans a pragmatic and innovative approach to life. This attitude was often contrasted with the stance of Europeans, thought to be mired in tradition and incapable of change.

In this context, Benjamin Franklin was unquestionably America's first great ingenious Yankee individualist and technology enthusiast. He believed, as did some of Europe's leading contemporary philosophers, that the universe itself could best be understood as a vast piece of machinery that had been put in motion by a distant God. This mechanistic viewpoint permeated every aspect of life and thought. The utilitarian economists who followed this line of thinking, for example, believed society should be based on a mathematically determined formula of "the greatest good for the greatest number" and that human happiness could be reduced to a measurable quantity.

In his *Autobiography*, Franklin spelled out in detail his attempt to achieve moral perfection in his own life. He approached this challenge by outlining 13 cardinal virtues in a characteristically systematic and mechanical manner, and proceed-ing to practice each of them for one week. He began with Temperance. Although he readily admitted his eventual failure to carry out his plan, Franklin viewed it as a useful exercise in self-knowledge. He had found it especially difficult to attain the virtue of Humility.

In Franklin's own time, science was viewed primarily as a philosophical and theoretical pursuit. This tone was set by the British Royal Society, whose members generally disparaged practical applications of scientific thought. Franklin, by contrast, was interested in both the theoretical and the applied sciences. Famous for his experiments with electricity, he was also one of the country's first notable inventors. Among his several inventions, his heating stove became a widely used household appliance, and his bifocal eyeglasses proved a boon to himself and many others. As a believer in empirical and applied science, Franklin once wrote:

> I would recommend it to you to employ your time rather in making experiments, than in making hypotheses and forming imaginary systems, which we are all apt to please ourselves with, till some experiment comes and unluckily destroys them.

If Benjamin Franklin was the godfather of American ingenuity, Thomas Jefferson was surely its father. Jefferson's home, Monticello, was itself the most visible product of his love of innovation and practical problem solving to improve everyday life. The house was filled with a variety of devices of Jefferson's invention, from specially made chairs to a dumbwaiter to convey food from the basement kitchen to the dining room.

Although Jefferson was knowledgeable in theoretical science, he was, like Franklin, equally adept in what was then known as "the useful arts." In an 1812 letter to a scientist he wrote:

> You know the just esteem which attached itself to Dr. Franklin's science, because he always endeavored to direct it to something useful in private life. The chemists have not been attentive enough to this. I have wished to see their science applied to domestic objects, to malting, for instance, brewing, making cider, to fermentation and distillation generally, to the making of bread, butter, cheese, soap, to the incubation of eggs, etc.

Central to Jefferson's ideal of the new Ameri-

The image of Benjamin Franklin's kite experiment with electricity became a popular icon of American science and invention.

can was the figure of the yeoman farmer, a small landholder leading a relatively self-sufficient life through agriculture. Jefferson assumed that this ideal farmer would be, like himself, well-enough educated to be an active and informed citizen and that he would take advantage of scientific and technological advances in all areas of life, particularly in agriculture. Like most contemporary Americans who were seeing the beginnings of the industrial revolution in this country, Jefferson assumed that technological change would bring progress and that change would be beneficial.

A rather different view of early 19th-century America came out of the travels and writings of the French intellectual Alexis de Tocqueville, whose book *Democracy in America* was first published in 1835. No single book can be taken as the final word on a subject so huge as the life of an entire society, but Tocqueville's work has been a touchstone for many generations of historians and social critics. His observations have served both then and now as a provocative and enlightening critique of the American scene.

Tocqueville commented on a wide variety of American characteristics, but his chapter "Why the Americans Are More Addicted to Practical than to Theoretical Science" is most pertinent here. He noted that:

> In America, the purely practical part of science is admirably understood, and careful attention is paid to the theoretical portion which is immediately requisite to application. On this head, the Americans always display a clear, free, original, and inventive power of mind.

However, reflecting the European viewpoint, he stated his opinion that Americans "carry to excess" this emphasis on practicality.

Tocqueville also argued that there was a link among the factors of social mobility, the pursuit of wealth in a democracy, and the emphasis on practical science, i.e., technology:

> As they are always dissatisfied with the position which they occupy, and are always free to leave it, they think of nothing but the means of changing their fortune, or increasing it. To minds thus predisposed, every new method which leads by a shorter road to wealth, every machine which spares labor, every instrument which diminishes the cost of production, every discovery which facilitates pleasures or augments them, seems to be the grandest effort of the human intellect.

In his concern that this attitude could be both destructive and beneficial, Tocqueville was perhaps ahead of his time.

In the 19th century, the emerging definition of American values based on individualism, pragmatism, and doing things for oneself was shared, sometimes inadvertently, by intellectuals as well as by engineers and entrepreneurs. It was reinforced, for instance, by writers like Ralph Waldo Emerson, whose essay titled "Self-Reliance" was a widely read and highly regarded statement of American self-perception. It is said to have been one of Henry Ford's favorite pieces of reading.

Emerson stated in the strongest terms his belief in the integrity and power of the individual:

> The secret of fortune is joy in our hands. Welcome evermore to gods and men is the self-helping man. For him all doors are flung wide; him all tongues greet, all honors crown, all eyes follow with desire.

In the very same essay, however, Emerson cautioned:

> The arts and inventions of each period are only its costume and do not invigorate men. The harm of the improved machinery may compensate its good.

Later in the century, Mark Twain, in his satirical novel *A Connecticut Yankee in King Arthur's Court* (1889) depicted just such a situation. The story is built on the premise that an ingenious, pragmatic, and enthusiastic American is transported back to the days of King Arthur. Describing himself, the Yankee says:

> Why, I could make anything a body wanted—anything in the world, it didn't make any difference what; and if there wasn't any quick new-fangled way to make a thing, I could invent one—and do it as easy as rolling off a log.

Appalled to find himself in an environment lacking the comforts and conveniences of modern life, he immediately sets out to industrialize and modernize the country. This premise gives Twain a magnificent opportunity to satirize both forms of naïveté: the pretensions of the courtly society and the ambitions of the American. The essential conflict is between the medieval society, based on magic and mystery, and the modern man's world, based on a scientific and materialistic viewpoint. Twain's story ends with a tragic battle between thousands of knights led by Merlin, and a small band of the Yankee's followers. In the closing scenes 25,000 knights lie dead, and the ingenious technical contrivances of the Yankee are dynamited by their creator. The American technologist has devastated the society he set out to improve.

Despite Twain's cautionary tale, the end of the 19th century found the stage set for an era of almost unqualified enthusiasm for technology. In America, as elsewhere, the dominant Victorian belief in the progress of the human race seemed vindicated by the marvels of technology. From Edison's electric light to the wonders of the automobile and beyond, most Americans seemed ready to embrace technology in almost every form. The 20th century saw the flowering of this enthusiasm in unbridled consumerism, fueled by the production, promotion, and acquisition of goods to make life more comfortable, convenient, and pleasurable.

Long ago, Emerson had warned that "Things are in the saddle, And ride mankind." However, it is only in the past several decades that a larger number of Americans have begun to think seriously about the costs as well as the benefits of technology. If the physical environment is endangered by consumerist technology, there may be a global price to pay. This perception represents a

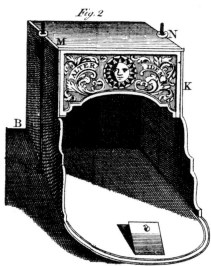

Benjamin Franklin's famous stove, designed just before he began his electrical experiments, proved an efficient source of warmth.

Thomas Jefferson's bed, built into the wall, permitted access to both bedroom and sitting room. The combination lounge chair and writing table was another invention illustrating his ingenuity.

dramatic shift from a relatively complacent acceptance of technology to a more questioning, indeed ambivalent, attitude.

The following essays, in no way intended as a definitive or comprehensive account of the enthusiasm for technology in 20th-century America, are offered as a series of explorations into largely uncharted territories of a vast cultural landscape. Addressing a variety of topics, they are linked by a common interest in the implications of America's love affair with technology.

At least two issues that recur in the essays highlight the complexity of understanding 20th-century attitudes toward technology. One of these is the relationship between technology and gender. With relatively few exceptions, it is evident that men have consistently created and maintained hegemony in the realm of technology. The scientific unraveling of the mysteries of heredity has yet to uncover a "technology gene" in male DNA, and it seems safe to presume that interest in, and aptitude for, things mechanical is a culturally learned trait.

Implicit in Joseph J. Corn's wide-ranging essay on technical literature, this modern bias is at least in part rooted in the centuries-old traditions of male-dominated artisanry, crafts, trades, and tech-

nical knowledge. Corn's analysis of the development of technical writing shows how the various genres of this literature have evolved to meet the needs of diverse audiences. His discussion of home repair books and magazines, as well as technical texts for children, seems to underscore the masculine bias.

Pursuing this theme further, Carroll Pursell thoughtfully demonstrates that the early 20th-century gender bias is strong and explicit enough to constitute a self-conscious effort to create a masculine world of technology exclusively for boys and men. In addition to the text and imagery found in the technical magazines, the boy heroes of numerous juvenile books, like the Tom Swift or The Young Engineers series, reinforced the male technological ideal. Finally, new toys like Erector Sets and Lincoln Logs helped to point boys toward the technical world.

There is little doubt that technology, like sports, has often served the purposes of male bonding, self-identity, and exclusion of women. This remains largely true on both the professional and amateur levels. While women are entering the medical and legal professions in growing numbers, the engineering professions continue to be dominated by men. Similarly, hard-core computer enthusiasts tend to be male. Any woman who ventures into a computer shop or a high-fidelity audio equipment store risks condescension from male sales clerks mouthing technical jargon. In her incisive essay, Susan J. Douglas explores the nature of such masculine obsessions as manifested in the development of early radio, hi-fi audio, and FM radio. Part of Douglas's thesis is that these pursuits provided socially acceptable outlets for masculine energy in a society that called increasingly for technical mastery rather than physical strength.

A second recurrent concern in the essays is the relationship between amateur and professional enthusiasm for technology. The amateur enthusiasm of ham radio operators as described by Susan Douglas was a form of recreation or play, in the sense that boys and men experimented with and tested the limits of the new toy. Her description of the exuberant early days of wireless in the 1910s and 1920s is a compelling portrait of amateur technologists. She also shows how such playfulness sometimes turned into mischief or outright troublemaking, using technology to subvert the social or technical order. A contemporary example might be found in adolescent computer hackers who create virus programs to damage large information networks.

Corn's outline of the genres of technical writing shows us how the rapidly increasing specialization of technology has been reflected in a widening gap between literature for professionals or serious amateurs on the one hand, and literature for the mass-consumer audience on the other. The ever-increasing profusion and diversity of special interest technical magazines is another indicator. Against this trend, just a few magazines such as *Popular Mechanics* have continued to appeal to a broad readership with a general interest in technology.

Nowhere is the importance of the amateur enthusiast for, and practitioner of, technology more evident than in the 90-year appeal of *Popular Mechanics* magazine. Mary L. Seelhorst has contributed a carefully detailed survey of persistent and changing directions and themes in that publication's long history. Like all popular publications, the magazine has served as both cultural

beacon and cultural mirror. As a beacon, it keeps its readers up to date with news of present and future developments in technology. As a mirror, it reflects for its readers an idealized view of themselves and confirms their membership in a community of others with like interests. The contents of the magazine tell us a great deal about its readers as, for instance, a self-defined community of do-it-yourselfers. My reminiscence of a 1950s do-it-yourself household serves as one example of a family devoted to, and influenced by, *Popular Mechanics*.

The relationship between amateurs and professionals is also dealt with in Robert C. Post's skillful essay on the American fascination with speed. In this realm, he points out, the line between the amateurs and the professionals is often invisible. Both are driven by the enthusiasm itself, by the desire to go farther or faster no matter what the cost. The thrill of the technology becomes a challenge and an end in itself. Post convincingly conveys the excitement inherent in the endless testing of technical possibilities. From street hot-rodding to the almost bizarre pursuit of the land speed record, devotees are consumed by the challenge.

The concluding essay, Michael L. Smith's provocative commentary on the post-World War II nuclear scene, pushes the professional-amateur split further in addressing the radical difference between atomic technology and all preceding technologies. The potential of nuclear power had first been demonstrated by the horrendous destruction wrought by the atomic bombs dropped on Hiroshima and Nagasaki in America's effort to bring an end to the war without invading Japan. In the aftermath, a stunned society tried to comprehend and domesticate this new technology, to stress its potential for positive contributions to the welfare of humanity. As Smith shows, the officially sanctioned projections of the uses of nuclear energy sometimes reached the absurd in fantasies of reshaping the earth itself by digging canals and harbors with atomic explosions. Technology was now Big Science, and its proponents needed to reassure a nervous public that the march of progress would continue unabated.

Despite these efforts, nuclear power has continued to serve as one of the focal points of public fear of the dangers of technology. As technology becomes more complex and esoteric, it becomes correspondingly more difficult for most lay persons to understand. In the age of computerization, the public increasingly experiences technology by buying it and using it without being able to change it or tinker with it.

The inability to obtain mastery over the machine may lead to a sense that technology is out of control, that it has indeed become an end in itself. This perception of technology as a Juggernaut that would crush everything in its path was defined at least as early as the beginnings of the industrial revolution in England. The poet William Blake decried the "dark Satanic mills" for blighting the rural countryside and transforming an agrarian way of life into the often desperate way of life of factory workers and tenement dwellers.

As we see in this book, in America the popular response to technology has been for the most part enthusiastic and optimistic. At the same time, there has always been an undercurrent of concern, at least in some quarters, about some of its side effects. Public debate about the long-range implications of our romance with technology is likely to continue and grow. But there is no reason to think that it will lessen the fascination of many Americans with the continuing evolution of technology.

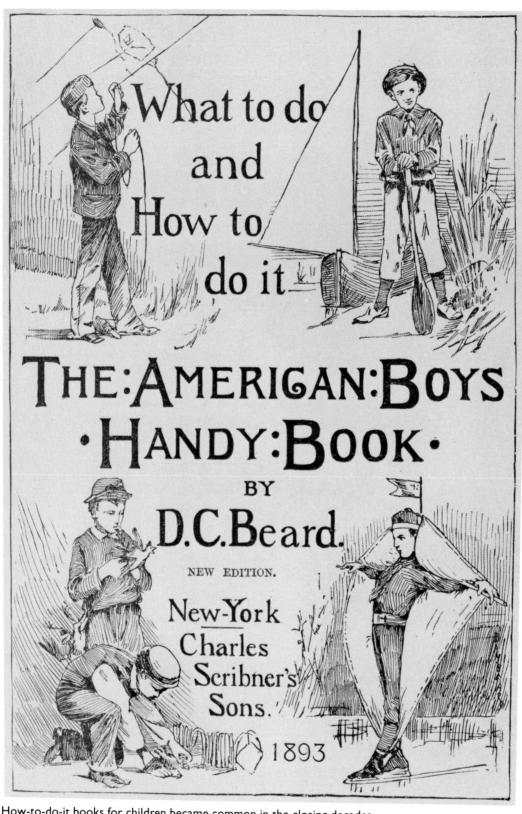

What to do
and
How to
do it

THE:AMERICAN:BOYS
·HANDY:BOOK·

BY
D.C.Beard.

NEW EDITION.

New-York
Charles
Scribner's
Sons.

1893

How-to-do-it books for children became common in the closing decades
of the last century. The author of this one was Daniel C. Beard, a founder
of the Boy Scout movement in America, and a prolific author of instructional
texts. Although being "handy" had long been valued, in the turn-of-the-
century era of urbanization many people encouraged handiness in the
young more as a means for building character than for economic or
vocational purposes. Scribner's also published *The American Girls Handy Book*
by Lina and Adelia Beard.

Educating the Enthusiast: Print and the Popularization of Technical Knowledge

SINCE THE CIVIL WAR, AN AVALANCHE

OF PUBLICATIONS ON TECHNICAL

SUBJECTS HAS INUNDATED AMERICAN

SOCIETY. RANGING FROM FACTORY

INSTRUCTIONS FOR BUYERS OF

MACHINES TO SELF-HELP MANUALS

FOR THOSE WHO WORK WITH

MACHINES, TO PERIODICALS

FOR HOME MECHANICS AND HOBBYISTS,

THIS LITERATURE HAS BEEN PRINTED

IN MASSIVE QUANTITIES

AND HAS BEEN WIDELY CONSULTED.

JOSEPH J. CORN

We live in an age of "popular mechanics." Virtually everybody today knows *something* about complex machines and technical devices—enough at least to own and operate cars, televisions, microwaves, telephones, and a host of other appliances common around their homes and workplaces. To be sure, few people would be able to repair such devices, and fewer still could build one from scratch. Compared to earlier eras, however, when only a few artisans or engineers had *any* involvement with complex technologies, knowledge about mechanics has become much more widespread over the last hundred or so years. Information about how machines are made, how they work, and how to keep them running has also become much more readily obtainable. Home mechanics wanting to install their own garage door opener or repair their VCR need do little more than read the manufacturer's instructions, or the book *How to Keep Your VCR Alive: VCR Repair for the Total Klutz,* to have a good chance of succeeding.

Part of what makes many Americans "mechanics," in the sense I am using the term, is their historically accumulated experience with technology. Over the last century or so, millions of us have used lawnmowers, typewriters, flush toilets, sewing and washing machines, phonographs and parlor organs, electric buzzers, burglar and fire alarms, typewriters and dictaphones, bicycles and automobiles, to mention just a few of the hundreds of devices familiar today. These machines have functioned as object lessons in the power of human artifice, demonstrating technology's ability to save labor, enlarge comfort, entertain, or augment and extend human abilities. Technology has also taught Americans another lesson: complexity. In coping with these new devices, in learning to exploit their potentials and adjust to their foibles, people have learned something about mechanics. To be sure, experience has seldom taught all that one has needed to know; increasingly, people have turned to more formal instruction, specifically to the printed word.

Our age of popular mechanics thus has been the era of popular technical literature as well. Since the Civil War, an avalanche of publications on technical subjects has inundated American society. Ranging from factory instructions for buyers of machines to self-help manuals for those who work with machines, to periodicals for home mechanics and hobbyists, this literature has been printed in massive quantities and has been widely consulted. While the "best sellers" of this genre are not studied in university English departments and seldom read straight through like novels, they have nonetheless played a very important role in modern life, one virtually ignored by historians. Without this literature, our complex machine-based civilization would hardly be possible.

Technical literature has a long history. Ancient writers such as Vitruvius, a Roman engineer and architect, offered advice in manuscript books on constructing buildings and designing weapons, yet only small numbers of people could read them and the high cost of the books—hand copied by scribes—constrained the spread of knowledge. The invention of printing by moveable metal type in the 15th century and the ability to print copper plate engravings broadened access to printed materials and stimulated the dissemination of knowledge about technology and other subjects demanding illustration. Yet not until the latter half of the 19th century did social and technical factors

combine to thrust the subject of mechanics before a mass audience.

A veritable information revolution took place, spurred by technical advances in the printing and distribution of texts. Mechanical paper making and the use of rotary steam presses helped lower the costs of producing books, and the telegraph and railroad facilitated the nationwide distribution and sale of reading matter. In the latter decades of the 19th century, innovations in the mechanical reproduction of images, which made possible the inexpensive inclusion first of lithographs and then of photographic images so essential to understanding machinery, further supported the popularization process.

Contributing to the increased supply of technical literature as well, was the growing tendency of participants in technical activities to organize themselves for regular exchanges of information. Engineers and inventors, for example, along with physicians, historians, and economists, founded numerous professional societies in the post-Civil War decades, as did various groups of tradesmen. All recognized the increased importance of keeping up with the growing quantities of new knowledge in their fields and of sharing such information with like-minded specialists. They usually did so through publication of proceedings or journals. In some fields, such as steam engineering or plumbing, government licensing of practitioners through examinations provided one more spur to the production of popular technical texts and to learning through reading.

Exemplifying the same impulse toward organization and specialization in the 1890s, technical publishing itself crystallized as a field. Publishers like McGraw-Hill and Norman Henley, both of New York City, built businesses either exclusively or substantially around the publication and distribution of technical titles. Correspondence schools, an educational innovation at the time, also added to the supply of technical literature. Led by the International Correspondence School of Scranton, Pennsylvania, these institutions issued textbooks and primers on subjects ranging from automobiles to electricity to plumbing.

On the demand side, the United States of the

By the turn of the century, employers increasingly sought workers with specific technical skills. The acquisition of these skills often required book learning, and this became accessible to millions via correspondence courses and self-study. This 1907 advertisement was placed in *Popular Mechanics* by the International Correspondence School of Scranton, Pennsylvania.

Here's the Opportunity
Are You the Man?

If an employer should say, "I want a man for an important position," would you be the right man?

late 19th century offered an almost insatiable market for popular technical reading. Widespread fascination with, and an eagerness to adopt, labor-saving technology explain some of this demand, but also significant was the relative absence in the United States of the strong tradition of apprenticeship and artisanal training that existed in Europe. American workers, less intensively trained and less inclined to stay with one trade for a lifetime, often compensated for their lack of formal training—or embarked on a new job path—via self-study and reading.

Demand also owed something to the country's large English speaking population and its high rate of literacy, both factors further augmented in the late 19th and early 20th centuries by the spreading ideal of universal schooling. As a result, the American publishing market could support a rich variety of specialized technical books and periodicals that countries with smaller, more linguistically diverse, and less literate populations could not hope to emulate.

By the turn of the century, then, the relentless development and diffusion of new technologies, processes themselves dependent on increasing supplies of printed knowledge, in turn catalyzed further demand for printed information about technology. Buyers and sellers of machines, along with owners and users, adjusters and repairers, all had new needs for information. Retailers and wholesalers wanting to know what products were being manufactured and what distinguished each particular model and brand turned to trade publications, advertisements, and manufacturers' solicitations. On the other side of the sales counter, consumers also sought printed guides to the new world of complex devices and supplemented word-of-mouth information with reading of their own. Having purchased a technical commodity, consumers looked to printed texts to help them adjust, maintain, and operate their complex mechanical possessions. A growing cadre of specialized repairmen emerged to perform the more difficult tasks and they too demanded ever more detailed information in the form of parts catalogs, service bulletins, and shop manuals.

Sink installations represented the entering wedge of high technology in the American home of the late 19th and early 20th centuries.

Even a relatively simple technology like indoor plumbing can illuminate this dynamic symbiosis between texts and technics. Although there was nothing inordinately complex about flush toilets and running water systems, it was not until the second half of the 19th century that such devices slowly began to supplant the traditional outdoor privy and well. As late as 1869, *The American*

Woman's Home, a best-selling guidebook on domestic science by Catherine Beecher and her novelist sister Harriet Beecher Stowe, omitted any discussion of running water or toilets, so rare were such features in the nation's homes. They said nothing about outhouses or wells either, but those were artifacts so well known and so unchanging over time that writers had never given them much attention. By the end of the century, however, the situation had changed dramatically.

Increasingly, water under high pressure was delivered to kitchen and bathroom taps via the new steam pumping engines that towns and cities were installing in their waterworks. New types of pipe and fixtures and new techniques of installation were necessary to move this pressurized water to and through houses because of the greater likelihood of damaging leakage. The recently invented flush toilet shifted an essential, if potentially noxious, human activity indoors, further complicating the technical infrastructure of dwellings. Scientists' discovery, about this time, that microscopic organisms can cause disease, underscored the necessity of keeping rigidly separate the two water systems that now were becoming part of every home and workplace: the sewage system, linked to toilets, and the fresh water system, tied to sink and bath.

A husband's ingenuity, coupled with a gas-powered stationary engine, mechanized one of woman's most arduous tasks—doing the Monday wash.

Popular technical literature not only responded to these changes in domestic technology but served to promote and guide them. By 1892 at least 17 different titles on plumbing and sanitation were in print, and it was no longer possible to write a book about domestic science without devoting significant attention to those subjects. Trade and professional journals also conveyed commentary and advice for the emerging professional communities engaged in plumbing and sanitation. The speed with which new knowledge, linked to new technology, grew in the plumbing field is captured by a comment in an 1899 treatise: "If the plumber of twenty years ago had fallen asleep and awakened only today," observed the author, himself a former plumber, "he would find himself without a trade."

Most trades and many other activities were being similarly transformed by a combination of new technologies and technical texts. Some of the inventions—automobiles, radios, and, later, computers—inspired more textual fallout than others, like lawn mowers or vacuum cleaners, about which few authors wrote entire books. Yet what stands out is the degree to which modern technology has been so thoroughly explicated by, embedded in, and, in a basic sense, constituted in people's minds by popular technical texts. To further explore this phenomenon, let us consider now the major genres of technical literature and their evolution.

The Modern How-To Book

The oldest technical texts, in the tradition of Vitruvius alluded to above, are individually authored books of advice that offer recipes for common processes or products, such as making soap or tanning animal hides, or that explain how to build or improve some artifact or device. With time, more specialized technical texts appeared, a movement spurred on by the erosion, in the 17th and 18th centuries, of the constraints which traditional guilds had once placed on their members against divulging trade "secrets." Once free of regulations, tradesmen increasingly took up their pens to tell outsiders how to perform some skilled technical task or work with a particular kind of machine or device. Such texts, authored by persons with hands-on experience, were the lineal descendants of ancient works like those of Vitruvius and the ancestors of modern how-to books.

The intention to explain and popularize a technical subject, however, did not automatically translate into an ability to do so. Many technical texts in the 19th century, although explicitly addressed to beginners, assumed a reader already knowledgeable about the rudiments of the field. The *Practical House Carpenter*, for example, published in 1830 by Asher Benjamin, the well-known American architect and proponent of the Greek Revival style, promised contents that were "methodised and arranged in such a simple, plain, and comprehensive manner as to be easily understood." Yet the section of the book on "Carpentry" freely employs a variety of terms without definition—mortise, tenon, joist, purloins, truss, and tie beams—thereby rendering the book less likely to be "easily understood" by people without prior experience in a workshop or on a construction site. A "glossary of architectural terms" appears in the book, but defines none of these words.

In the same vein, although Benjamin's work contained "sixty-four large quarto copper plates," distinguishing it from the ordinary 19th-century how-to work with no or a few illustrations, none identified the elements of a structure alluded to in the text. The images in Benjamin also assumed considerable preliminary knowledge on the part of the reader. Drawn in the traditional orthographic projection favored by engineers and architects—as plan views and elevations—most of them lack the perspective readers were likely to know from paintings, engravings, and other more popular visual representations.

Writing that seldom defines terms and assumes an insider's familiarity with a field characterizes Benjamin's book and most of the others in the growing output of 19th-century how-to literature. I call this style of writing the artisanal literary tradition. It grew out of a situation in which tradesmen learned their craft through an oral culture rooted in the workshop. There, apprentices learned virtually everything, from the names of tools and the parts of machines to work techniques, by listening to, observing, and emulating somebody more skilled than they. In such an environment,

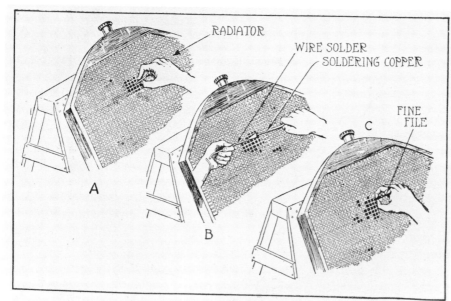

RADIATOR · WIRE SOLDER · SOLDERING COPPER · FINE FILE

This step-by-step, sequenced illustration of how to repair radiators is a good example of the growing sophistication of technical graphic art in the 20th century.

Fig. 179.—Steps in Repairing Leaky Radiator.

rank beginners acquired enough verbal and visual information to start being useful very quickly; mastery of the field took much longer. Authors of 19th-century how-to books had acquired their technical knowledge this way, and understandably wrote as if they were addressing apprentices already knowledgeable about the rudiments of the field. They seldom described or analyzed what they considered as commonplace in the workshop. As a result, their books are not "easily understood" by neophytes but only by those already familiar with the basics of joinery, millwork, steam engineering, or other technical fields.

By the second decade of the 20th century, however, a pedagogical revolution in such literature took place, led by what one author called "the discovery of the novice." This discovery represented the flip side of the growing value placed on "expertise," especially in scientific and technical fields. As the gulf between what experts knew and what ordinary people knew widened, and as educators and others reflected on how to close such gaps through instruction and schooling, the so-called "novice" stood out as never before. But while the line between novice and expert, amateur and professional, became more defined, a growing number of novices sought to acquire the specialized information that would enable them to cross it. At the same time, greater numbers of experts began to author texts that would enable novices to acquire more expertise.

In the 1920s, the targeting of the novice led to the proliferation of how-to-do-it texts written in a different voice for a differently conceived reader. These works seem fresh, and to our late 20th-century sensibilities, they appear eminently "user friendly." Moreover, the best of them *really are* "easily understood," even by neophytes. The authors took care to define terms, lay out procedures for performing work in step-by-step fashion, and offer other aids in the form of easy-to-follow construction plans or how-to-do-it illustrations.

Many of these new texts came from the automotive field, where complex and trouble-prone machines catalyzed a great demand for easily digestible information. Victor Page's *Automobile Repairing Made Easy*, which first appeared in 1916, exemplified this new type of volume. Although the title page labels the work with the somewhat foreboding designation "treatise," the 1,060-page book was advertised as being "written by an Expert and is written so you can understand it." The author approached his subject "in a practical, non-technical manner" and claimed that the book would "prove of equal value to the chauffeur, owner and general mechanic."

Page, like other modern how-to authors, went beyond merely what to do; he tried to bring readers into the *process* of work, describing various ways of performing a task, evaluating the merits of each, and warning the reader what to avoid and how to recognize pitfalls. Along the way, he divulged such useful bits of shop lore as the usefulness of "common yellow laundry soap" to stop a leak where a fuel feed pipe attaches to a carburetor. Marking the book as modern and revolutionary was the inclusion of "over 1000 specially made engravings."

Along with the traditional kind of engineering projections found in earlier books, Pagé's book employs many new types of printed images. Some illustrate the tools used to adjust and repair automobiles, a significant departure from 19th-century practice by which tools received little mention and almost never any graphic treatment. Such knowledge about tools was precisely what one acquired during the first few weeks on a job and thus was not something that a writer in the artisanal literary tradition was likely to mention, let alone illustrate. Another innovation is the use of detail drawings,

such as a depiction of "Points in the Valve Operating Mechanism That Demand Attention," which focuses a novice's attention on a particular problem area. On this and other drawings, the relevant parts or features are also clearly labeled; reproductions of photographs in the book have superimposed labels (or arrows) to identify specific features or parts.

The most significant graphic innovation evident in Pagé's and similar texts was the "hands-on image," a drawing or photograph of a worker's hands performing some task or using a particular tool. Because it is helpful in learning any physical task to watch an expert carry it out, hands-on images conveyed information that illustrations without a human presence never could. In the 19th century, it had been assumed there was no

Nineteenth-century trade magazines often published "shop kinks," hints, and advice for artisans who worked with their hands. In the 20th century, *Popular Mechanics* offered homemakers "home kinks" in its "Solving Home Problems" column. The format of the column clearly signals a rigid division of labor according to gender.

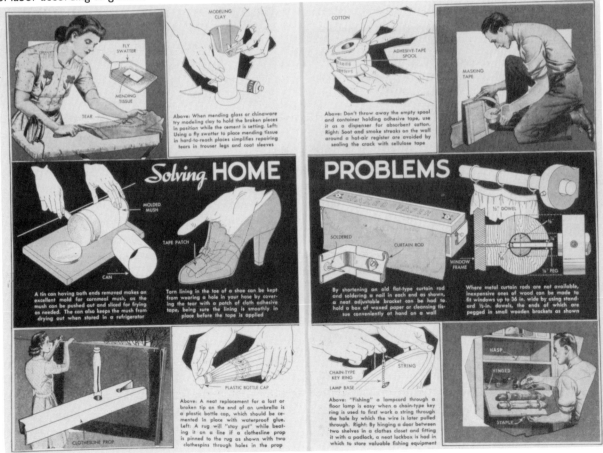

way to have "the eyes and fingers of the experts accompany the essay," as one writer put it, a point still true even with today's video how-to cassettes. Yet like videos, the hands-on image provided a useful approximation. By seeing the expert's fingers, perhaps in the act of adjusting the screws on a carburetor, a novice receives considerable help in emulating such work. The reader studying such an image is also, in effect, looking at the carburetor through the expert's eyes. He can therefore more easily orient himself to it and separate what is important from the cacophony of material complexity surrounding it.

Finally, the representations of expert hands in a text symbolized the willingness of author and publisher to admit novices into their world of expertise. Scrutinizing such images while reading the accompanying texts, deploying tools, and putting his or her own hands to work, the reader of the modern how-to text joins a dispersed, print-mediated variant of the 19th-century artisanal workshop community. Although the illustrated technical text could never wholly equal the more dynamic, intimate, and detailed first-hand instruction available to beginners in the social setting of the shop as a force for democratizing and expanding access to technical know-how, it has been immensely influential.

Its legacy surrounds us everywhere. How-to books today are available to teach one how to build everything from an aeroplane to a yurt, or how to transform oneself into a Chinese cook or an avocational machinist. These how-to narratives, whether in book or magazine format, have become the materialist bibles of the modern era; above all they have made mechanics popular.

Manufacturers' Publications

Even persons who never purchase how-to books are familiar with the now ubiquitous owner's manual, the printed instructions which accompany virtually all technical devices. There is a long history to this relationship of texts to machines, although exactly when it began is not clear. By the 1820s and 1830s, however, manufacturers of inexpensive, mass-produced clocks were already textualizing their products this way. Affixed to the back of a clock or inside it, buyers would find a label instructing them how to ready the clock for operation by attaching the pendulum and how to set the time. Similarly, by the latter half of the century, purchasers of agricultural machinery, sewing machines, bicycles, and a host of other things often found themselves poring over accompanying technical instructions. And, as was true of early how-to books, much of this printed material, often technical, brief, and lacking in clarity and specificity, must have left its readers quite perplexed.

Nineteenth-century buyers of machines, for example, found themselves confronting printed instructions like the four-page booklet provided by Adriance, Platt & Co. of Poughkeepsie, New York, for its 1867 "Buckeye Mower." Typical in its lack of detail, this brochure functioned as an advertisement, an instruction manual, and parts catalog. The cover featured a crude woodcut illustration of the Buckeye Mower "Mowing on Side Hill, and Running over a Rock." The intention of the illustration seems to have been to advertise the Buckeye's capabilities under conditions that would have led other brands of mowers to break down, rather than to offer useful information about the operation of the Buckeye.

The second page of the booklet, titled, "Putting Up and Using the Buckeye Mower," constituted the germ of what eventually would evolve into the owner's manual. Unlike more recent complex machines, however, the "Buckeye" had to be assembled by the buyer of (or possibly the sales agent for) the machine, and so the manufacturer's booklet listed six steps to follow in performing that task. Yet the booklet provided no illustration showing how the various parts fit together! To be sure, the brief instructions mentioned most of the parts by name, but if one had no prior knowledge of mowers and did not know what particular parts were called, assembly might be extremely difficult. Only by studying the "Price-List of Extra Parts of Buckeye Mower No. 2" on page three, wherein each part was listed and numbered, with the numbers keyed to illustrations on page four, could an owner match the name and appearance of each part. How each part related to the others, however, was not explained.

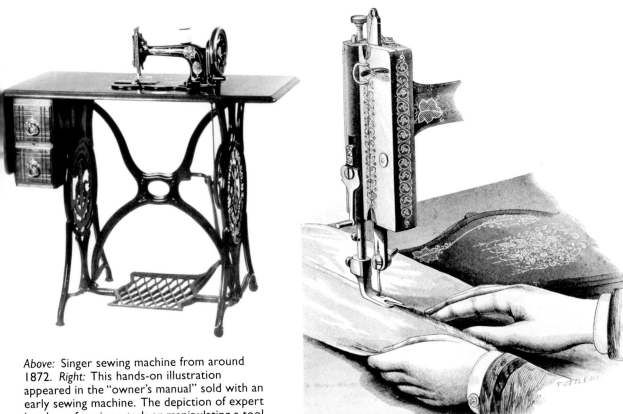

Above: Singer sewing machine from around 1872. *Right:* This hands-on illustration appeared in the "owner's manual" sold with an early sewing machine. The depiction of expert hands performing a task or manipulating a tool or machine, although common in today's how-to literature, was unusual in the 19th century.

Felling. (See Page 15.)

Whoever wrote the Buckeye pamphlet may have assumed that his farmer-readers were relatively familiar with tools and agricultural technology, and since mechanization had been altering farm life for a generation or more, the assumption might well have been justified. By contrast, housewives in the same period were relatively unfamiliar with machinery, which may explain why some of the most thorough and, to our eyes, "modern" technical writing went into the production of instruction books to accompany the sale of sewing machines, the most complex devices marketed to women in that period.

The "Household" Sewing Machine Instructor, for example, published in the 1870s by the Providence, Rhode Island, manufacturer of the machine, differed markedly in degree of helpfulness from the Buckeye Mower pamphlet. Instead of a single page of text, it contained 26, and instead of 2 illustrations it contained 17, each carefully engraved and most full page, including a "transparent view in perspective" (today known as a "cutaway" view). On this view, the various parts of the sewing ma-

chine were also numbered, with matching numbers employed throughout the text, so that readers could more easily learn the name of a part, master the machine, and follow the instructions on how to operate it. Many of these illustrations were "hands-on images" of the sort we have already talked about. Indeed, the extensive reliance on this feature of modern how-to pedagogy is, for a 19th-century text, quite unusual.

As already suggested, a clue as to why this 1870s manual appears so modern lies in the nature of the illustrated expert hands: they belong to a woman. While women had for a century been operating machinery in textile mills, only with the sewing machine did middle-class women confront mechanization in their own homes. Such women had little prior and direct experience with gears, bearings, pulleys, and other mechanical components; as a group they were more likely than farmers to be novices when it came to machinery. Authors of sewing machine instruction books, I would suggest, fully realized this situation and thereby developed more user-friendly instructional

Although the book industry does not collect statistics on technical best sellers, surely John Muir's 1969 work belongs on such a list. His publication, addressing the "compleat idiot" of his subtitle, went through 8 editions and 18 printings in less than a decade.

texts. Indeed, although the *Instructor* was published anonymously, as were virtually all owner's manuals, a woman may well have been its author, in view of the frequency with which women wrote independently published how-to books about sewing.

Female how-to-do-it writing in the 19th century generally tended to assume an audience of neophytes, quite unlike texts written by men. As to why this should have been true, we can only speculate. Seldom were women graduates of artisanal workshops, and thus they may have approached the writing of technical texts less concerned about restating the obvious than men were. Women were also society's traditional experts in communicating with beginners, that is, with infants and children. Such experiences may have freed them from worrying about being perceived as talking down to their audience when they wrote for publication.

In any event, by the 20th century, "instructors" and other texts published by manufacturers—some written in modern voices and truly understandable to beginners and others providing only stingy and elliptical guidance—had become common features of the information environment of most Americans. These texts rapidly became more specialized. Twentieth-century consumers infrequently encountered booklets like the 1867 "Buckeye Mower" publication that functioned as advertisement, assembly and operating manual, and spare parts catalog all at once. Automobiles, with their thousands of parts and great complexity, led the move toward specialization. Instead of comprising a few pages in the back of a booklet, spare parts listings for cars grew to massive volumes indexing myriad parts, many slightly different depending on make, model, and year. And in the automotive world, the owner's manual, itself a new invention in the final decades of the 19th century, had by the mid-1920s given birth to a more specialized and comprehensive offspring, the so-called "shop manual." These volumes offered a guide to all of the adjustment, repair, and replacement operations one might perform on a car. All the technical literature offered by manufacturers embedded those machines ever more deeply in a network of texts essential to their continuing use. Consulting parts catalogs, shop manuals, or other manufacturer's literature has become a necessary—if for many people tedious and confusing—part of modern life.

Technical Magazines

Another genre of technical literature, the modern technical magazine, also appeared in the 19th century. One version of the genre was the trade journal, which proliferated after the Civil War, addressing the technical and business concerns of electricians, plumbers, railroad engineers, automobile dealers, and many other groups. By "giving the machine a voice," as an editor of the *American Machinist* put it, tradesmen could in effect listen to and learn from technology without being in the presence of the actual machine. As with science, where printed discourse had been shared through the medium of scientific journals since the latter 1600s, now technology also possessed texts that bound its practitioners together. Technologists could therefore learn through published commentary rather than, as in the past, only through the oral culture of the workshop. Readers of trade journals and other technical magazines were not simply learning passively; many of them wrote letters to the editor, offering advice or criticizing articles they had read. Because such letters were commonly published, a national and even international dialogue emerged and came to exercise a broad educational influence. Such exchanges helped standardize technical practice over large geographic areas, fostered new ways of doing things, and further stimulated innovation.

While the trade magazines addressed readers primarily in their roles as producers, by the early years of the 20th century a second type of technical magazine catered to those whose relationship to new technologies was more that of consumer. These "consumer periodicals," as we may call them, dealt with a range of technical questions and subjects of interest to men and women in their day-to-day lives, in purchasing equipment for the home, for the office, or for leisure pursuits, or in using, improving, maintaining, or repairing their dwellings or possessions.

Popular Mechanics, one of the earliest examples of the consumer magazine, typifies the genre. First published in 1902, the magazine by 1910 had a circulation of a quarter million and by 1950 was on a list of the top 30 mass-circulation magazines. In its first decade or so, the magazine seems to have been more of a regular trade journal. Aimed at an audience of skilled shopmen, it offered advice and ideas to those in the metal trades, steam engineering fields, and mechanical occupations generally. In the 1910s, however, *Popular Mechanics* shifted its emphasis to become a source of information for technically minded homeowners and homemakers. Instead of articles dealing with repairing railroad car-wheel bearings, changing belting on power machinery, or guying telephone poles against winds—all of interest to tradesmen or those in mechanical occupations—the magazine now focused on work around the home, especially for men who, in a newly popularized word, were supposed to be "handy." In its new guise, *Popular Mechanics* instructed readers how to construct

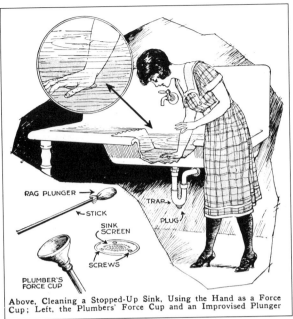

Above, Cleaning a Stopped-Up Sink, Using the Hand as a Force Cup; Left, the Plumbers' Force Cup and an Improvised Plunger

This do-it-yourself tip from a 1925 issue of *Popular Mechanics* suggests using either a store-bought device, the plumber's force cup or plunger, or a folk remedy, namely, the palm of one's hand, as an "improvised plunger."

built-in furniture for the dining room, repair cracked china, wire a door bell or garage door opener, and carry out any of the thousand other maintenance and improvement projects necessitated or suggested by the modern, technology-filled home. Besides tutoring the growing number of do-it-yourself enthusiasts, the magazine also catered to those who chose such craft-oriented hobbies as, say, building ham radio sets or models of locomotives or sailboats.

The number and variety of other magazines that emulated *Popular Mechanics* in covering the burgeoning world of do-it-yourself and technical hobbies grew rapidly. As with so many areas of modern life, specialization dominates this story. Magazines came into being for almost every imaginable consumer, hobbyist, and do-it-yourself taste, as is evident from these titles and their initial year of publication: *Motor Boating* (1907), *Popular Radio* (1929), *Model Railroader* (1934), and *Consumer Reports* (1936). Even in the age of television, when millions do not read any magazines, the list of new publications serving specialized avocational and hobby interests continues to grow. In recent years, the expanding market in nostalgia, for collecting, restoring, or just reading about old cars, phono-

The night before Christmas children dreamed of new toys, while parents often translated such dreams into reality by following the instructions of do-it-yourself publications like *Popular Mechanics*.

graphs, typewriters, and other kinds of machinery, has given rise to dozens of new technical magazines, such as *Live Steam* (1976; steam-powered engines and models) or *Cars and Parts* (1956; auto restoration and collecting).

What all of these periodicals have in common, starting with *Popular Mechanics*, is their dependence on a relatively affluent, leisured, and often suburban audience, a population of consumers. Starting in the late 19th century, increasing numbers of Americans moved to the suburbs where many men were likely to take up a technical "home trade" such as carpentry or radio. Frequently employed in sedentary, white collar jobs, some took up such activities for exercise or relaxation. Others were forced to such pursuits simply because it was

difficult to find a professional nearby to perform minor work on their homes. Still others simply wanted to learn more about the machines on which they were increasingly dependent—how to use and care for them—or wanted to occupy their spare time productively with some form of handicraft.

Whatever their motivations, and whoever the readers of the expanding menu of the new technically oriented consumer magazines were, they were more likely to be relatively ignorant about tools, processes, and technology generally, than readers of trade publications. To successfully communicate with these new consumer-readers, magazines tended, in the words of the early *Popular Mechanics* slogan, to be "written so you can understand it." What this meant was eschewing technical language, spelling out in detail the sequence of steps to be taken in making or fixing something, describing problems that might arise along the way while offering advice on how to avoid them, and illustrating profusely the entire work process.

To promote understanding, *Popular Mechanics* widely employed some of the techniques we have already discussed, including frequent use of hands-on images. The magazine also used other aids that have since become common in communicating about crafts to neophytes: a "bill of materials," for example, that enumerates the number and dimensions of pieces of lumber or other materials, the types and quantities of hardware, and the additional supplies needed to complete a specific how-to-do-it project; occasionally, the magazine even estimated the dollar cost and time necessary to make something. Articles of a "back to basics" nature with such titles as "How to Use Twist Drills" or "Files and How to Use Them," also became common. Perhaps the ultimate novice-oriented feature, introduced by the magazine in the 1940s, was "From the Wood Pile." Using no text, save for an occasional caption, and occupying only a single page, this feature showed readers *graphically* how to build simple items such as a wastebasket or knick-knack shelf using scraps of wood likely to be found around the house.

In a sense, "From the Wood Pile" encapsulated the psycho-economic logic of the novice-oriented, mass-market periodical: simplify the how-to-do-it message as much as possible so as to reach the largest audience (that is, the *most* ill-informed) possible. If a text could be, to paraphrase the *Popular Mechanics* slogan, "written so a child could understand it," which might mean writing without words, so much the better. Indeed, almost from the beginning *Popular Mechanics* acknowledged this logic and, through its department called the "Boy Mechanic," sought to reach the youth (at least *male* youth) market. In 1913, Popular Mechanics Press republished the columns from the department in a book of the same name. Periodically into the 1950s, the Press offered new editions of *The Boy Mechanic*, all of which, because they specifically addressed children, represented a wholly new phenomenon in popular technical publishing.

Technical Texts for Children

Children have been the ultimate technological enthusiasts, fathoming new technologies with an ease marveled at by their elders. This now familiar phenomenon first appeared during the turn-of-the-century decades as telephones, streetcars, aeroplanes and other technical marvels dazzled adults but simultaneously left them feeling ill at ease and ignorant. In response, many sought to better prepare their children for the future in which such technologies would be dominant. Increasingly, they shaped children's playtime and education around new technologies, especially the play of young boys, who were traditionally envisioned as the engineers and technicians of the future. As boys encountered such technical toys as erector sets or electric trains, they also confronted technical texts, just as their parents did in dealing with new machines. For the first time in history, a technical literature addressed just to boys came into being.

These texts fall into our two already familiar genres: how-to books and various manufacturers' publications. How-to volumes for boys or young men usually covered a wide range of craft projects rather than a single trade or field, like *The Boy Mechanic* volumes already mentioned. One of the pioneers of the genre, and also a founder of the Boy Scouts of America, was Daniel Carter Beard. His *American Boys Handy Book* first appeared in 1882 and was followed by similar works, including *The Outdoor Handy Book* (1893) and *The Jack of All Trades, or, New Ideas for American Boys* (1900). As their titles suggest, these volumes emphasized outdoor projects for young woodsmen: how to make sleds and boats, how to camp and fish, and how to collect and preserve bird eggs and nests. Beard and other contemporary authors of children's how-to texts also addressed more technical topics, such as constructing flying whirligigs and kites, miniature sawmills, or pinhole cameras.

The *Boy Mechanic* volumes published by Popular Mechanics Press also mixed outdoor-oriented subjects such as "How to Make a Flint Arrowhead" or "Camps and How to Build Them" with such high-technology ones as "How to Build a Glider" or "Home Made Telephone Transmitter." While this blend of pioneer nostalgia for the recently closed frontier and celebration of technologies that

To the luckiest men in the world
A FATHER AND SON WHO ARE PALS

This 1933 advertisement for Lionel trains would have yielded limited results. In the depths of the Depression only the truly prosperous could afford this standard gauge train set and girder bridge whose combined cost approached that of an automobile.

were transforming the world might seem paradoxical or contradictory, the two kinds of articles in fact taught boys similar lessons. Both offered guidance on how to read diagrams and how to think spatially. Similarly, they imparted knowledge about tool usage and, most significantly, the importance of following detailed, sequential directions. Thus youngsters, whether inspired to construct a tepee or a steam engine, in both cases were arguably fitting themselves for a society in which print-based interactions with tools and objects were ever more essential.

At the leading edge of technically informative play, however, were certain toys that utilized up-to-date technology, such as electric trains, which were also embedded in a web of youth-directed texts. Early on, the catalogs of the industry leader in the field, the Lionel Corporation, pitched the firm's toy train "outfits" to boys rather than their parents. By the late 1910s their increasingly lavish, multicolored publications fueled the technical daydreams of legions of youngsters. Lionel catalogs became a youth-oriented variant on the Sears Roebuck wish book: a text to dream over and a primer for the joys of future possession. Yet the catalogs offered more than merely a spur to consumerism. They served as important tutors to boys on a range of technical matters. Even when they never obtained the train of their desires, there was much for them to learn by reading the detailed specifications governing Lionel construction or the catalog's advice on building a miniature layout. One learned, for example, about the different applications of sheet, rolled, and pressed steel; about the construction, uses, and care of batteries (used for operating electric train outfits, as many homes were still not wired for electricity); and about the behavior of that mysterious force itself, electricity. One also picked up much information about railroading.

In boys' relationships to electric trains, mediated as they were by the ever popular Lionel catalog, we glimpse in microcosm the larger development about which we have been concerned in this essay. Whether they were young or old, whether playing with toys or using full-scale machines, people's interactions with complex technologies over the last hundred or so years have increasingly been shaped by printed technical literature. As with the Lionel catalogs, buyers and users of all kinds of technologies have become accustomed to turning to printed sources for guidance on purchasing and operation. Similarly, mechanics routinely consult texts—shop manuals or parts catalogs—to put right devices that malfunction or have broken down. All of this technical literature has proved a source not only of enthusiasm but also of knowledge. Most importantly, it has had significant consequences in our society.

In a country that reveres individual choice and celebrates the possibilities of personal transformation, how-to-do-it literature has found an eager audience. Such literature not only mirrors those values but has also contributed to occupational mobility and achievement, although the precise effects here would be impossible to gauge. Through print, anybody who can read can take steps toward mastering a new craft or trade. Thus one who has never built so much as a shelf, picks up a book and, with little further help, builds a house; somebody who didn't know an escapement from an escarpment, purchases *Clock Repairing as a Hobby* and soon has the old grandfather clock in the hallway chiming regularly again. The dizzying choice of avocational or home trades accessible through popularly available technical texts, further resonates with what may be called the American jack-of-all-trades ethos which, although revering experts and expertise, still celebrates individuals who can do many things adequately and who, in a nutshell, are "handy."

Because American artisans and tradesmen historically lacked the skills and training common to European or Asian artisans, how-to-do-it literature has played a crucial part in technical education in the United States. Over the years, many "professional" tradesmen acquired their skills through regimens of self-study and reading, sometimes assisted by correspondence schools, sometimes not. Similarly, many amateurs converted their hobbies or avocational trades into full-blown occupations through reading mixed with experience, first earning a few dollars fixing clocks for neighbors and then hanging out a shingle as a professional clock repairer. This movement from amateur to professional standing in mechanical trades has been aided by the American tendency to view hobbies, in the words of historian Stephen Gelber, as "a job you cannot lose" and to talk about leisure activities using the same vocabularies of profit, moral benefits, and achievements associated with remunerative work. The line between amateur and professional is further blurred by the many former hobbyists who have prospered in such technical fields as model building, auto restoration, and computer programing. Their success is in significant part attributable to the abundance of technical literature and ease of access to it.

Besides fostering occupational change and mobility, technical texts have influenced the transformation of the American built environment. The explosion of handbooks, treatises, catalogs, and manuals on plumbing in the late 19th century, for example, not only enabled plumbers and architects to design and build new homes and offices, but also facilitated the monumental task of gradually retrofitting many thousands of older structures with pipes and waste lines so as to bring in running water and modern toilet facilities. Although it is probably impossible to trace the influence of particular texts in this process, the mere fact that such books were published and sold testifies to their importance. Catalogs depicting modern bathrooms and kitchens whetted people's appetites for those improvements, while the increased supply of easy-to-understand how-to-do-it texts encouraged homeowners and others to undertake the retrofitting and/or remodeling of bathrooms and kitchens. Such projects, in turn, stimulated the desire for yet more information and advice.

In such ways, the interactive symbiosis between changing technology and printed texts has been both extensive and fruitful in the United States. The now familiar process of textualizing technics continues. Our enthusiasm for technology, flowing ultimately from the artifacts themselves, derives as well from their surrogates in print. Popular technical literature has helped make technology popular.

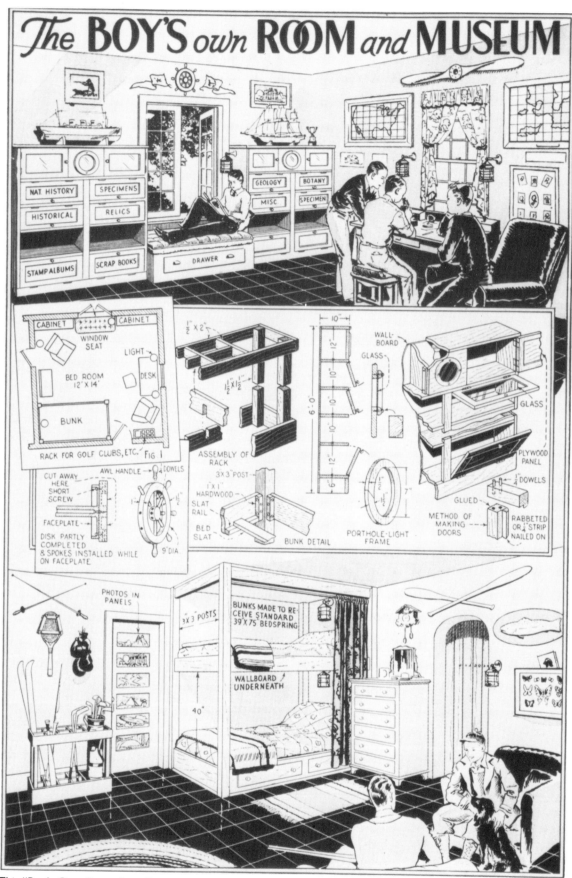

This "Boy's Own Room," as illustrated in *Popular Mechanics*, is a museum of social as well as natural history. Masculine bonding with peers and Father take place in a context of science and technology.

The Long Summer of Boy Engineering

HISTORIANS HAVE CHARACTERIZED THE
GENDERING OF AMERICA IN THE WAKE
OF THE INDUSTRIAL REVOLUTION AS
THE CREATION OF "SEPARATE SPHERES."
PRIVATE SPACE, ESPECIALLY AT HOME,
WAS THE DOMAIN OF WOMEN, WHERE
THEY HAD SPECIAL RESPONSIBILITIES
AND POWERS. PUBLIC SPACE, POLITICS,
COMMERCE, WAR, ENGINEERING, WAS
THE SPECIAL PLACE OF MEN,
WHICH THEY NOT ONLY CONTROLLED
BUT FROM WHICH WOMEN WERE EXCLUDED.

CARROLL PURSELL

The April 1933 issue of *Popular Mechanics* featured what it called "The Boy's Own Room and Museum." On a page, separated by the expected detailed and dimensioned sketches of cabinet parts, two scenes pictured the room of what must have been, for many families, the Ideal Boy. In the top sketch our young hero sits reading in a window seat flanked by cupboards clearly marked Nat. History, Historical, Stamp Albums, Specimens, Relics, Scrap Books, Geology, Botany, and Misc. Atop the cabinets are model ships (sailing vessel and luxury liner) and a trophy cup. On the wall are pictures of a hunting scene and another ship, maps, a ship's steering wheel, and an aeroplane propeller. His neatly dressed friends are gathered at another table, looking through a microscope.

In the bottom scene we see the other side of the room with bunk beds, a butterfly collection, sports equipment, a stuffed fish, and on one chair the boy himself dressed ready for the woods, and in another chair his father with a shotgun on his lap. The boy is patting his dog.

As so often happens, the picture tells more than it realized and more than it meant to. The first decades of the 20th century were ones in which the American middle class had particular cause to worry about its sons. Their fathers had moved from the small towns and farms of the nation to large cities where they were typically white-collar work-

35

ers in the growing corporate bureaucracies of the country. They had worked hard in their own childhoods, and their economic contributions to the family income had been welcomed and often even necessary. They had had their characters formed (or so they thought) by hard work and the imposed reality of need. Importantly, they had had their developing masculinity nurtured and shaped by their fathers, alongside whom they often worked. The next generation of boys however, cut off at once from the discipline of real labor, the joys and lessons of field and stream, and the constant presence of a father's masculine influence, seemed dangerously at risk.

Historians have characterized the gendering of America in the wake of the industrial revolution as the creation of "Separate Spheres." Private space, especially the home, was the domain of women, where they had special responsibilities and powers. Public space, politics, commerce, war, engineering, was the special place of men, which they not only controlled but from which women were excluded. Of course the definitions of public and private were never completely clear, and were frequently contested. Over the decades women had worked hard to expand their domain of competence into areas clearly "public," such as the social problems of poverty, education, and sanitation, which could be legitimized as natural extensions of caring for the home and family. By the turn of the century, the rise of modern science and technology, however, was recapturing such areas for men. As science rather than morality became the measure of public policy (and even of good housekeeping!), those with scientific training (almost exclusively men) were clearly privileged. This new "Wise-man's Burden" gave scientifically and technically trained men the responsibility for leadership and control to stem the rising tide of women and people of color at home and abroad.

"The Boy's Own Room and Museum" is a powerful response to that fear. In his city or suburban room, the Ideal Boy is surrounded by many wholesome, constructive and evocative op-

The technology and violence of war, translated into toys and games, were well calculated to make "men" of these boys. In a few short years, they, in their turn, would be called upon to inflict and suffer the horrors of World War II.

This Warship Buckles Amidship and Slowly Sinks to the Bottom of the Bathtub When Its Vital Spot is Struck by a Swift Toy Torpedo

portunities for activity and contemplation. Father is linked particularly to the sport of hunting, but the other sports equipment, references to the sea, and decorative maps all conjure up a masculine domain of struggle and adventure where male bonding can take place. No touch of the feminine intrudes. As much as the references to hunting and sports, the many artifacts of modern science and technology also point to the manly pursuits of a man's world. In countless books, stories, advertising images and, increasingly, movies, the masculine ideal of America was defined as the engineer.

Outside the room, in the real world, organizations such as the Boy Scouts of America (founded 1910) were being established to build the characters of boys through the organized activities of camping and handicrafts. By 1981 some 65.5 million American men and boys had passed through scouting. Perhaps even more, however, had enjoyed that staggering outpouring of science and technology toys, kits, models, and books that flooded the American market in the first half of this century.

Traditional ships' models were joined, within months after the successful first flight of the Wright brothers, by model aeroplanes. In 1913 a national organization was being formed to tie together local clubs dedicated to the building and flying of model planes. In "Boys as Airplane Modelers," published in *Illustrated World*, November 1913, Edward I. Pratt claimed that the boys, "while profiting in most ways by the experience of the real bird

Build this 36-inch Scale Model of the
SS UNITED STATES
page 182

In this issue...
CHRISTMAS GIFT SECTION
PLYMOUTH RESTYLES FOR '53
115 IDEAS for the CRAFTSMAN

Male bonding through technology was a persistent theme of the literature for boys. It is unclear, however, whether a model ocean liner brought out more of the man in the boy, or vice versa.

men and scientific model builders, work out their own planes from actual experience in flying the machines." An advertisement published after World War II made explicit the argument of building character and showed its persistence. The Comet Industries Corp. published *Meet Cap-* *tain Comet "Jet Ace,"* to introduce boys "to the fascinating world of aviation." Their slogan was "Model Building Builds Model Boys," and in an age of "togetherness" that feared "juvenile delinquents," the Comet Corp. had reassuring words for concerned parents, "model building directs

Transportation Has Always Been a Problem to Man, and Here Are Three of the Latest Modes of Travel for the Youngsters

By ALFRED ALBELLI

VERY often the careers of great men in various scientific fields have found their first inspiration, according to Ferdinand Strauss, head of a large toy factory, in the playthings which amused and fascinated them in childhood.

He goes so far as to assert that there is no inventor who cannot recall the influence, slight and subtle as it might have been, which tinkering with toys has had on his bent to inquire into what composes their mechanical elements, and in eventually shaping his life's course.

"Many a youngster," said Mr. Strauss, "is started early on a career through the encouragement he receives as he takes his first toy apart to see how it works, and then tries to put it together again.

"The desire to know what a toy is made of and why, what it is supposed to do and how it does it, teaches the child many things. History is full of examples of

men who received the incentive for their future life work from their playthings.

"Everyone knows that Thomas A. Edison spent all of his playing time with tools. Neither did the Wright brothers invent the airplane by accident. During their childhood days they continuously played with balloons. They would inflate different-sized balloons and float them at various heights to observe how the wind blew them and see which floated highest.

"Carl F. Akeley, the inventor and big-game hunter, also spent his play hours as a boy with various types of small implements. The gift of a folding pocket footrule, he has told me, gave him a big thrill, and with his tools he made toys and useful articles for the house. In later years, during his famous exploits in the jungle, hundreds of miles from any source of supplies, this ability of construction stood him in good stead. His fun with a tool chest

as a boy now turned to great practical worth in facing the crises of travel and shelter.

"Another great man in whom the influence of toys is traceable is Dr. George Ellery Hale, of the Mount Wilson observatory. He is one of the world's foremost astronomers. Among his favorite toys were engines, yachts, and other machines that afforded opportunity for mechanical construction.

The Advance in Aeronautics within the Last Few Years Has Made Its Effect Felt in Toyland, and the Ingenious Models and Toy Airplanes Have Proved a Source of Almost as Much Amusement to Grown-Ups as to Children; at the Right, an Airplane Model That Appears to Have Caught the Fancy of Someone Long Since Removed from Childhood; Below, a Little Girl and Her Family of Forty Dolls, Including Almost Every Kind, from the Sawdust-Filled Rag One to the Dainty China Dolls That Talk and Cry and Walk

From the architect Frank Lloyd Wright to the astronomer George Ellery Hale, a host of successful scientists, architects, and engineers testified to the influence of construction toys in boyhood.

youthful energy into constructive channels; in the home, it becomes a unifying influence, as father and son work together in building faithful replicas of famous airplanes."

In the early years of the automobile age, when statistics on family car ownership were skyrocketing, it is not surprising that vehicles scaled down for boys were enormously popular. For aspiring Soap Box Derby entrants, *Popular Mechanics* in its October 1923 issue carried an article by George A. Luers with plans for "Building a Four-Passenger Hand Car." Carrying out the plan would, it was predicted, "provide a great deal of healthy exercise and fun for the boys

building it." For those who preferred to buy rather than make their cars, the Dail Steel Products Co. in Lansing, Michigan, "conceived the idea that the most outstanding mode of transportation in a modern child's mind was an automobile." In 1924 it published *Wolverine Juvenile Vehicles*, promoting "the latest in juvenile automobile design." Where the four-passenger hand car had to be pushed, the 1924 Wolverine was pedaled by its driver. The 1921 Auto-Wheel Coaster was marketed in ways reminiscent of adult vehicles. "If Tom has an 'Auto-Wheel,' you can be sure that Dick and Harry will want one too!," exclaimed the company. To make sure of this it had

enrolled 25,000 boys in Auto-Wheel Clubs, all of whom received copies of the magazine, the *Auto-Wheel Spokes-man,* "a lively little publication full of good live stuff that every boy likes to read." The company's 1921 publication *The New Models* presented a scheme whereby it would "help boys to buy their 'cars' and often [would] actually advance club funds for this purpose." It had been only two years since the General Motors Acceptance Corporation had become a pioneer in credit selling.

The chemistry set had been available since the mid-19th century, but now was joined by a number of construction sets including Tinker-Toys and Lincoln

Logs, the latter designed in 1917 and patented in 1920 by John Lloyd Wright, the son of Frank Lloyd Wright, and an architect as well. Such toys, made up of a large number of simple, unassembled identical pieces, were well designed for the new techniques of mass production. The back pages of each issue of *Popular Mechanics* were peppered with small ads directed at boys: for Build-A-Motors, model ships (the *Sovereign of the Seas*), "Cleveland-Designed Flying Models" of aeroplanes, a home microphone, a small gasoline engine for aeroplanes, a crystal radio receiver, and a booklet titled the "Boy Amateur Electrician" (10 cents). These ads all mingled with others for "The Ventrilo" (Boys! Throw Your Voice), stage money, trusses, and a scheme for easily learning to play the saxophone.

When it came to providing kits and toys, however, nobody could approach A. C. Gilbert. Like Teddy Roosevelt, whom he resembled in many ways, Gilbert had been sickly as a child but made up for it through sports, setting a world's record in the high jump while in college. He earned a medical degree at Yale, but never practiced, pursuing instead the manufacture first of magic and then of science and technology sets. He informed his industrial career with a muscular notion of what boys liked and wanted.

In his 1920 catalog *Boy Engineering,* he wrote:

I feel that every boy should be trained for leadership. It is only the bright-eyed, red-blooded boy who has learned things, done things, dared things beyond the reach of most boys who will find the way open to really big achievements.... My toys are toys for the live-wire boy, who likes lots of fun and at the same time wants to do some of the big engineering things—things that are real—things that are genuine.

The catalog offered a wide range of kits to tempt the boy market. Gilbert's obituary in the *New York Times,* January 25, 1961, quoted him as saying, "I've remained a boy at heart and only introduced items that appealed to me, I figured they would appeal to all boys."

In 1920, his line of wares included a toy tractor ("every wide-awake boy knows what wonders the Tractor has accomplished, and what a tremendous aid it is in the great farming districts of the West. You boys want to see

Thomas Edison was still alive when this kit was offered for sale to boys. In his old age, Edison even cooperated in a nationwide search to find a boy genius to follow in his steps.

how these up-to-the-minute machines work") and a machine gun ("if there ever was a real live-wire toy for the red-blooded boys, this is it. Say, you can have more genuine sport with this machine gun than anything I know of. It's the real thing"). Kits included those for hydraulic and pneumatic engineering, magnetism, sound, meteorology, machine design, signals and electricity.

It was his Erector set, however, that made Gilbert famous. Invented in 1914, shortly after the British Meccano set which it strongly resembled, the Erector

set was made up of the soon familiar strips of stamped metal fastened together by small nuts and bolts. "Hello Boys!," read an ad in the December 1933 issue of *Parents' Magazine,* "You Build Like a Real Engineer When You Build With Erector." For the Christmas trade that year Gilbert produced his "big illustrated 'Look-Em-Over' Book" which contained information on, and an entry blank for, his annual Erector model contest. Boys were urged to invent some new machine built of Erector parts, and submit it for prizes ranging up to

a trip to "the Panama Canal, or Boulder Dam or the Empire State Building or any other engineering project in the United States." They were also invited to tune to Gilbert's Sunday evening broadcasts (perhaps on a crystal set radio they had put together themselves) of "Engineering Thrills," "true stories about real engineers and their hair-raising adventures in digging the Panama Canal, and in building bridges and skyscrapers."

The theme of engineering adventure early found its way into books. Tom Swift, the boy inven-

Wonderful Construction Blocks of Real Stone

A whole toyland of fun for boys and girls— and an industrial art education as well— are to be had with these scientifically made blocks of real stone— in three natural colors. They are the finest construction toy a boy or girl could have. There are no parts to rust or to get hurt with, none to shrink or swell or fall apart. At all times, in all climates, these blocks are unchanging, correct, clean and everlasting.

Castles, battlements, bridges, forts, arches and towers rise as by magic under the child's hand. Each block is made architecturally exact and mathematically accurate. They are easy to put up and easier to take down—no pegs or screws or nuts or bolts—and with one set of Anchor Blocks you can build an almost endless number of models. Books of designs furnished free with every set.

RICHTER'S
ANCHOR BLOCKS

Each block fits perfectly to the other blocks so that it is possible to buy additional pieces or regular supplementary boxes at any time to add to the set so that larger models or several models can be built. Anchor Blocks develop initiative and the constructive side of a boy or girl's nature.

They broaden the mind and the imagination. They lead to higher ambitions. Buy a set now for that boy or girl of yours, or get one for Christmas. Prices 50c to $5.00. Fortress sets for building modern forts at $1.50, $3, $6, and $12.00.

Write for illustrated catalog today. Just mail us the coupon below, writing your name and address plainly or send money order if you wish to order now.

F. AD. RICHTER & CO. 74 Washington Street Dept. 202 NEW YORK CITY

A touch of the tomboy was always allowed. Sister joins in the constructive fun, but mother takes her reposeful ease. Father, of course, supervises. Brother might well go on to a career in technology but that is unlikely to be the case for his sister.

MECCANO

THE TOY THAT MADE ENGINEERING FAMOUS

STEAM SHOVEL

The British Meccano was similar in origin, appeal, and success to Gilbert's Erector set. It is a reminder that neither engineering aspirations nor a masculine identification with technology was America's alone.

tor, appeared in 40 books between 1910 and 1940. Less well known is the work of H. Irving Hancock, creator of the *Young Engineers* series. Born in 1868, he listed himself as a chemist as well as an author, but appears to have taken no degree in that subject. He became a reporter for the *Boston Globe* in the late teens and served as a war correspondent in Cuba and the Philippines. Early in the century, he began to write the various books in his nine separate series for boys.

His series on the Young Engineers followed the adventures of Tom Reade and Harry Hazelton, described in *The Young Engineers in Mexico* as "a pair of young civil engineers who, through sheer grit, persistence and hard study . . . made themselves well known in their profession." His *Grammar School* series featured Tom, Harry, and four of their friends, who stayed together for the *High School* series as well, mostly distinguishing themselves in sports. Two of the friends then went off to West Point in the *West Point* series and two others to the navy in the *Annapolis* series. In his books, sports, war, and engineering all served as appropriate venues for developing and exercising the masculine virtues.

Significantly, a career in engi-neering allowed Tom and Harry to avoid the college experience of their other friends, and to go straight into an apprenticeship "in a local engineering office in their home town of Gridley. Then, with vastly more courage than training, Tom and Harry went forth into the world to stand or fall as engineers." *The Young Engineers in Colorado* found the boys building a railroad in the Rocky Mountains. *The Young Engineers in Arizona* found them again building a railroad, this time in the desert. *The Young Engineers in Nevada* told the story of their introduction to mining engineering, an activity that was continued in *The Young Engineers in Mexico.*

In all of these adventures, which, significantly, were set in the "Wild West," the boy engineers could act out the masculine virtues of courage, hard work, perseverance, honesty, and loyalty to friends. They used their initiative, struggled against nature (and human nature, in such persons as "Bad Pete, a braggart and scoundrel of the old school"), accomplished real things in the real world, and generally exercised a degree of personal autonomy that must have had a powerful appeal to boys just emerging from childhood, through adolescence to-ward manhood.

A generation later, *Popular Mechanics* brought out a series of "Adventures in Science" books, covering such subjects as chemistry, atomic energy, and electronics. One, titled *There's Adventure in Civil Engineering,* written by Neil P. Ruzic and published in 1958, aimed to capture the romance of engineering but made no effort to give the boy protagonists autonomy. Randy Morrow and his little brother Sam, travel the length of the Pan-American Highway in 1957 in the company and care of their father, known only as Mr. Morrow. Naïve but not unintelligent questions by Randy are answered by Mr. Morrow and a series of engineers they encounter along the route.

As the dust jacket explains:

They talk to lean clear-eyed engineers whose very calm confidence suggests high adventure in exciting places. The travelers see huge bridges being flung across mighty chasms, and mighty mountains blasted and gouged until they yield a place for man to move with machines.

It was hoped that "all mid-teen boys, and many girls, as well, will find in these pages that truly THERE'S ADVENTURE IN CIVIL ENGINEERING!." But the girls would not have found them-

There's Adventure in civil engineering

by Neil P. Ruzic

Randy Morrow finds adventure on the Pan-American Highway

ANOTHER POPULAR MECHANICS CAREER BOOK

The entirely masculine world of Randy Morrow celebrated the conquest of nature, the "opening up" of Latin America, and the engineer as carrier of manly virtue, economic development, civilization, and international amity.

selves in the book. As resolutely masculine as "The Boy's Own Room and Museum," the world of Randy Morrow is without women. The only female mentioned is a "white goddess" who is said to be the leader of a tribe of "dart-blowing headhunters" in the jungles of Central America, but she is dismissed as merely a native superstition.

It was altogether appropriate that Randy, Sam, and Mr. Morrow should look for evidence of American civil engineering adventure in the "underdeveloped" lands of Latin America. The Monroe Doctrine had early marked this area as peculiarly ours, and within the memory of people still living, Theodore Roosevelt, future president and most masculine of American role models, had led the charge up Cuba's San Juan Hill during the Spanish-American War. Roosevelt had also acquired the land for the Panama Canal ("we stole it fair and square," he was supposed to have said), which opened in 1915. In Randy's own time, President Harry S Truman's Point IV program was sending American technical and scientific know-how around the world to help prevent the triumph of communism in developing nations. Not for the last time, America felt good about itself and confident of its power and manliness as a result of successful adventures undertaken in foreign countries.

For half a century the ideal of the Boy Engineer was put forward in many guises and on many occasions. Since much of the driving force behind it was an anxiety for the "future of the race," it was designed to mold the character of middle-class boys into something approximating a model of American manhood. Not surprisingly, the masculine virtues admired turned out to be those associated in popular myth with the frontier: courage, ingenuity, hard work, and personal autonomy. Engineers especially, from the late 19th century on, were seen as most perfectly embodying this ideal. In the Old West and the new American overseas colonies, and still in Latin America, they sought to direct "the great sources of power in Nature for the use and convenience of man," as the British engineer Thomas Tredgold had put it a century earlier.

If anxiety about the character of American boys (and perhaps more broadly about the manliness of American men in general) provided a motive, the recent history of the nation suggested the means. The same processes of industrialization that were producing a generation of clerks and bureaucrats instead of pioneers and farmers were also creating a world where new technologies, often shaped by science, were restructuring both production and consumption. The desire to pour the old wine of masculinity into the new bottle of scientific technology proved overwhelmingly attractive.

The enthusiasm with which engineering toys were sold and used was palpable. Their ubiquity guaranteed that boys would absorb not only a working knowledge of the new technology of the 20th century but would develop the mental skills and attitudes to invent the next generation of improvements. Boys were urged to experiment, do real things, be creative. In *Meccano Prize Models: A Selection of the Models Which Were Awarded Prizes in the Meccano Competition 1914–15*, the Meccano Company, a British firm, claimed that its set really "does teach boys engineering. All the time they are building models they are acquiring knowledge which may some day prove of the greatest practical value to them."

An article by G. A. Nichols in the April 26, 1923, issue of *Printer's Ink* quotes Lincoln Logs inventor John Lloyd Wright as insisting that:

[A] real American boy with a keen brain is just about the smartest and most original thing alive. Precedent and custom mean nothing to him. He is bold and courageous in the execution of his ideas for this reason.

"Everything in life is a game," Gilbert once mused, "but the important thing is to win." With so many live-wire, red-blooded boys steeped in the fun and adventure of engineering, America was certain to be a winner.

Part of the attraction of technology for boys was that it never lagged far behind the state of the art. Within months of the establishment of the National Aeronautics and Space Administration (NASA), these boys were playing astronaut.

Parents built the 13 subassemblies in top secrecy. Right, punctured tin can projects "blips" on radar screen

Closed television circuit really works. Crew member in control compartment is visible on power-deck receiving screen. Picture is "transmitted" by three mirrors

Center right, brave crew of the X-3 arrives on the craters of the moon, sends back to earth the message "Mission accomplished!"

Space Ship, Junior

ON A SPECIAL launching pad near Syracuse, N. Y., the X-3 stands ready for blastoff. As the countdown crackles over the intercom, the crew makes ready for a new adventure in outer space.

The X-3 is perhaps the only space ship in the world made of plywood, nails, paint, gadgets left over from hobby kits and imagination. The ship was built by M/Sgt. and Mrs. F. S. Kalinowski (he's in the Air Force) for their three youngsters. Until it was completed, ready for its first flight to Mars, the ship was a top-secret project. The two parents spent 60 evenings of hush-hush work producing the ship's 13 subassemblies.

At last, in the dead of night, all the parts were smuggled out of the basement and the X-3 was quietly assembled. Soon after daybreak, the wide-eyed kids discovered that a spaceship had landed in their own backyard. ★ ★ ★

Crew makes preflight check before each blastoff. There are two working levels inside

Below right, lights of the "atomic pulsator" blink on and off, and space member informs rest of crew that reactor is operating

SEPTEMBER 1954

35 CENTS

POPULAR MECHANICS
MAGAZINE
WRITTEN SO YOU CAN UNDE

REG'D TRADE MARK, GREAT BRITAIN. No. 410426

Hi-Fi Is Sweeping the Country!
What's It All About?
—Page 106

Report on the
CADILLAC

HOW TO SHOOT
OVER DECOYS

A cover story in *Popular Mechanics* reports on the hi-fi craze that
swept the country in the late 1940s and early 1950s.

Audio Outlaws: Radio and Phonograph Enthusiasts

THE APPROPRIATION OF AUDIO
TECHNOLOGY IN THE 20TH CENTURY
BY BOYS AND MEN PROFOUNDLY
INFLUENCED TECHNICAL HISTORY AND
AMERICAN MASS CULTURE. THE TINKERING
OF THESE AUDIO OUTLAWS SET THE
STAGE FOR RADIO BROADCASTING,
REVOLUTIONIZED THE PHONOGRAPH
INDUSTRY, AND PIONEERED THE USE
OF A NEW FREQUENCY BAND, FM.

SUSAN J. DOUGLAS

Tinkering with machines, once the province of engineers, factory workers, and inventors, in the 20th century gave rise to a host of hobbies that often became obsessions for boys and men across America. Whether tinkering with cars, radios, or computers, young men could literally or figuratively transport themselves away from the confines of their homes and communities, and explore unfamiliar locales filled with new, sometimes forbidden, experiences and information. Often their uses of these technologies were defiant, challenging the ways in which entrenched corporations hoped their inventions would be used. But in the process of indulging in technological rebellion, these boys and young men often compelled the corporate world to reconceptualize how its machines fit into the marketplace. These subcultures of technical enthusiasts have, over the years, significantly reshaped America's cultural landscape.

This was certainly the case with radio, an invention which, from the beginning, seemed to serve as an irresistible lure for such young people. It was also the case with the hi-fi set. As will become clear, the often passionate oppositional uses of audio technology by boys and men produced technical fraternities whose uses of these inventions profoundly influenced mass culture. The first such appropriation began just after the turn of the century, with the emergence of amateur wireless operators eventually dubbed "hams." Later, in the 1940s and 1950s, some technically minded men were seized by another audio craze which led them to build their own hi-fi sets. Then, in the late 1960s and early 1970s, young men discovered FM radio—an invention that had lain dormant for 30 years—and used it to redefine radio broadcasting in America.

What all these technical subcultures had in common was that their use of audio technology deviated significantly from the expectations of the originators and producers of these inventions, as well as from the business interests that took them over. The degree of conscious defiance animating members of these groups varied, however. While a subgroup of the "hams" challenged the government's automatic appropriation of portions of the electromagnetic spectrum for military uses, and the hi-fi enthusiasts repudiated what they saw as technological complacency in the phonograph industry, the underground FM programmers attacked the entire political and cultural establishment as they saw it. These technologies allowed for—even invited—oppositional, anti-establishment uses primarily by white middle-class men and boys, who were expected, and eventually compelled, to integrate into institutional bureaucracies, yet who yearned to postpone such integration. Their use of these technologies allowed them to rebel. But it also provided them with critical technical expertise that would become valuable in the job market.

Why did audio technology prove such a magnet for young men? Why were these technical subcultures so avowedly masculine? Certainly women were excluded from even entertaining technical tinkering as a pastime, both by being barred entry to professions in science and technology and by being socialized into assuming a technophobic stance toward most machines, especially those related to electronics. But these various audio enthusiasms also allowed men to navigate conflicting messages about masculinity, and to gain access to realms often categorized as "feminine"—gossip and music—and to redefine them as male.

The stereotype of women's allegedly insatiable hunger for gossip had been given a new twist by the early 20th century when American popular culture featured countless caricatures of women's incessant, seemingly frivolous uses of another communications technology, the telephone. The constantly eavesdropping switchboard operator was another stereotype reinforcing the image of the telephone lines as a feminized communications network dedicated to chatter. Wireless allowed men to transcend this network, and to create their own in which they could commune with each other, free from the interventions of women.

Cultural contradictions about how men were to relate to music informed the technical activism of both the hi-fi enthusiasts and FM rebels. Although the composing, conducting, and performing of music had always been dominated by men, there was, simultaneously, an association for American men between the love of certain kinds of music—especially opera and classical music—and effeminacy. This connection had been exacerbated by the segregated education of middle-class boys and girls in the 19th century, when music became one of the "lady-like arts" that girls learned. And the strong strain of anti-intellectualism in American culture that regarded the "egghead" as effete contributed to the sometimes uneasy position of the musical devotée. For men who loved music but were eager to avoid such associations, technical tinkering was one way to resolve such contradictions. By the mid-1960s, when critics characterized the demise of 1950s rock'n'roll as the "feminization" of rock music, the stakes were quite clear. FM had to be used as a venue for reinvigorating rock music with that brand of male rebellion and virtuosity that had spawned rock'n'roll to begin with.

In all of these cases—wireless, hi-fi and FM—men with their own technical and social agendas appropriated still underdeveloped audio technology and pushed it to new levels of performance and new realms of application. Their oppositional activities exposed areas of corporate and technological myopia. The corporations managing these technologies had to respond to the innovations of these hobbyists, and did so by co-opting and taming outlaw practices to create huge new businesses.

The Miracle of Wireless Communication

On Sunday, November 3, 1907, the *New York Times Magazine* featured as its lead story an article titled "New Wonders with 'Wireless'—And by a Boy!" The youthful star of the article was Walter J. Willenborg, a previously unknown wireless experimenter and a student at Stevens Institute of Technology in Hoboken, New Jersey. A large oval portrait of Willenborg in the center of the page was surrounded by photographs of his home-built wireless station, which included transmitting and receiving equipment. The reporter described in excited detail all the messages he was privy to by listening in to "the ether" on Willenborg's headphones. Willenborg made such good copy that he was also featured in a 1908 issue of *St. Nicholas*, "An Illustrated Magazine for Young Folks."

Willenborg was one of the young men the press chose to represent the burgeoning number of nameless amateur operators in the country. Since 1899, when Guglielmo Marconi had first introduced his wireless telegraph to the United States during the America's Cup races, the prospect of sending telegraph messages through "the air" without wires had generated enormous excitement in newspapers, magazines, and the technical press. This excitement helped spark a new fad, and from 1906 onward, thousands of primarily white,

middle-class boys and men began to construct their own wireless stations in their bedrooms, attics, or garages.

Although they were to be found throughout the country, these amateur operators were most prevalent in urban areas, especially those with seaports. They hoped to listen in on messages sent by the navy, commercial ships, and shore stations, as well as to send Morse code messages back and forth to each other. They earned no money as operators and had no particular corporate or professional affiliation. For them, wireless was a hobby, one that required technical knowledge and skill. The technical fraternity these amateurs formed was exclusive. Working-class boys with neither the time nor the money to tinker with wireless could not participate as easily. Neither could girls or young women, for whom technical tinkering was considered a distinctly inappropriate pastime and technical mastery a distinctly unacceptable goal.

The amateurs' ingenuity in converting a motley assortment of electrical and metal castoffs into

Marconi's vision for wireless telegraphy, that it be used to send point-to-point messages in Morse code, was transformed by "ham" operators who used the device to pioneer the broadcasting of voice and music.

The Transmitter

PART ONE. WIRELESS

THE ELECTRO IMPORTING CO.
233 FULTON ST. 23 RUE HENRI MAUS
NEW YORK BRUSSELS
U.S.A. BELGIUM

This cover of an early how-to wireless manual shows a young "ham" at his station sending a message in Morse code.

working radio sets was quite impressive. With performance analogous to that of an expensive receiver now made available to them in the form of the inexpensive crystal detectors introduced in 1906, the amateurs were prepared to improvise the rest of the set. They had no choice before 1908, for very few companies sold equipment appropriate for home use. As the boom continued, however, children's books, wireless manuals, magazines, and even the Boy Scout manual offered diagrams and advice on radio construction.

In the hands of amateurs like Willenborg, all sorts of technical recycling took place. Discarded photography plates wrapped with foil served as condensers; cylindrical Quaker Oats containers wrapped with wires became tuning coils. One amateur recalled that he improvised a loudspeaker by rolling a newspaper into a tapered cone. Another inventor's apparatus was constructed ingeniously out of old cans, umbrella ribs, discarded bottles, and various other articles. The one component that

was too complicated for most amateurs to duplicate, and too expensive to buy, was the headphone set. Consequently, telephones began to vanish from public booths across America as the amateurs lifted the phones for their own stations.

By 1910, amateurs surpassed the U.S. Navy (the major governmental user of wireless) and the private wireless companies in numbers and, often, in the quality of the apparatus they owned. In 1911, *Electrical World* reported:

> The number of wireless plants erected purely for amusement and without even the intention of serious experimenting is very large. One can scarcely go through a village without seeing evidence of this kind of activity, and around any of our large cities meddlesome antennae can be counted by the score.

The *New York Times* estimated in 1912 that America had several hundred thousand active amateur operators. Even after passage of the Radio Act of 1912, which sought to regulate and stifle amateur activity in the air, the number of enthusiasts continued to grow. Between 1915 and 1916, the Commerce Department licensed 8,489 amateur stations, compared to fewer than 200 commercial shore stations. Estimates placed the number of unlicensed receiving stations at 150,000.

One characteristic seemed especially prevalent among these amateurs: their disdain for authority and their delight in using this new technology to flout it. While most amateurs used their equipment to gossip, trade technical information, share football or baseball scores, or compare homework, some were considerably more mischievous. The increased presence of amateurs in the airwaves, at a time when tuning was crude and interference was common, led to a struggle for control of the ether. This struggle especially pitted the more defiant amateurs against the U.S. Navy. Pretending to be military officials or commercial operators, they would dispatch naval vessels on all sorts of fabricated missions. Navy operators would receive emergency messages about a ship that was sinking off the coast. After hours of futile searching, the navy would hear the truth: the "foundering" ship had just arrived safely in port.

On the part of some, this was simple pranksterism, the sort of delinquency that is irre-

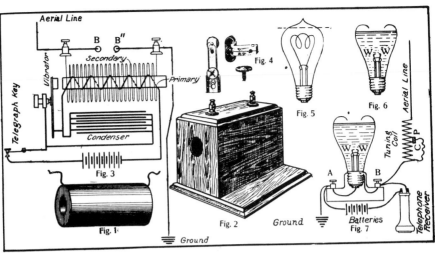

A *Popular Mechanics* article by George W. Richardson, "How to Make an Efficient Wireless Telegraph," provided step-by-step instructions on building a homemade set. *Below right:* An amateur station assembled at home, with the sending key to the left and tuning coil to the right.

How to Make An Efficient Wireless Telegraph

sistible when the target is distant and detection virtually impossible. But other amateurs had a more thoroughgoing critique of what they saw as an arbitrary usurpation of the airwaves by the state, and expressed their indignation by sending obscene messages to naval stations, and arguing extensively with naval operators over ownership of the ether. Military officials complained bitterly to Congress about what they regarded as etheric outlaws, and the more politically conscious amateurs responded by sending their own representatives to Washington to testify against military domination of the spectrum. The navy was hardly helped in this skirmish by the often romanticized portrayals of wireless operators in the popular press.

Increasingly, magazines, newspapers, and popular fiction celebrated the wireless dabbling of these young men. Fictitious Tom Swift, boy inventor, used radio to rescue people in distress, and by the 1920s there was an entire series of adventure books called *The Radio Boys*. Stories like the ones written about Willenborg captured the many attractions that wireless experimentation might hold for a young man. On a practical level, a successful wireless dabbler could make extra money from his pastime. He would have technical knowledge and skills few others possessed. He learned a code and he became an explorer. Through wireless, he entered an invisible, mysterious realm, somewhere above and beyond everyday life, where the rules for behavior couldn't be enforced—in fact, were not yet even established. He could participate in

contests of strength, power, and territory, by interfering with or interrupting other stations' messages, and win them without any risk or physical danger. In this realm, by mastering a new technology while letting his antisocial inclinations run loose, he could be, simultaneously, a boy and a man, a child and an adult. He could also straddle the more traditional definition of masculinity with its stress on physical activity and courage and the new definition with its emphasis on quite different characteristics.

The physical culture movement of the 1890s, the explosion in competitive sports, the revival of boxing, and the glorification of the strenuous life by President Theodore Roosevelt all equated true masculinity with the assertion and testing of physi-

GOVERNMENT OPERATES TALLEST RADIO TOWER

An executive order recently issued from Washington has placed the operation of the Tuckerton, N. J., radio station in the hands of the United States Navy Department, which will maintain it on terms of equality for all belligerent and neutral nations until the close of the European war. Since the ownership of the plant is in dispute—both French and German companies having made application for a government license—the government determined to take charge of it until the close of the present hostilities. Code messages will be received under strict censorship by naval officers. While the station will be open for commercial business, the only communications accepted in code will be those of foreign embassies. All other wireless stations capable of transmitting and receiving transatlantic messages are being operated by the owning companies under the supervision of the Navy Department, to insure that the neutrality of this country is not violated. The tower of the Tuckerton station, which was first described and illustrated in this magazine, is 825 feet in height, the tallest structure in America.

Amateur stations were temporarily shut down during World War I, when the U.S. Navy assumed control of radio transmitting in America. The navy gained access to very powerful stations such as the one in Tuckerton, New Jersey, initially built by a German firm.

cal strength. A new respect, even reverence, for man's "primitive" side was revealed in the success of Jack London's *Call of the Wild* and Edgar Rice Burroughs's *Tarzan*. Nor was it enough to be physically vigorous; men had to have forceful, commanding personalities as well. All of these traits, it was believed, were best cultivated through a physically more active life in which men were more directly in contact with nature.

At the same time, it was becoming clear that in the business world, physical strength mattered little: there the notion of physical combat was a metaphor for other kinds of confrontations. Increasingly, what landed a young man a good job, what gave him an edge in the race for success, was a combination of intelligence, education, and cer-

tain skills. The increase in high school enrollments, the growing popularity of adult education, and the self-improvement craze all attested to the new importance of education and specialized knowledge. Boys educated in both academic and corporate institutions learned that having a forceful personality was, in reality, often either unattainable or a liability. Despite the prevailing mythology, much of a man's life was spent indoors, in urban areas, in routinized, hierarchical bureaucracies, away from the enlivening and therapeutic tonic of the outdoor life.

For a growing subgroup of American middle-class boys, these tensions were resolved in mechanical and electrical tinkering. Trapped between the legacy of genteel 19th-century culture and the pull of the new primitivism of mass culture, many boys reclaimed a sense of mastery, indeed masculinity itself, through the control of technology. The boys lacking physical prowess could still triumph over nature if they controlled the right kind of machine. Playing with technology was, more than ever, glorified as a young man's game. And few inventions were more accessible to the young man than wireless telegraphy.

In a culture that was becoming more urbanized, and whose social networks were becoming increasingly fragmented, many strangers became friends through wireless. The fraternity that emerged developed the sense of fellowship characteristic of pioneers. Here in the ether, safe from superiors, sequestered from women, and free to abandon codes of politeness and civility, young men could act out their deep-seated need for interpersonal communication, for contact and a sense of community, for eavesdropping and for gossiping, while regarding all these needs as distinctly masculine.

A revolutionary social phenomenon was emerging. A large radio audience, whose attitude and involvement were unlike those of other, traditionally passive audiences, was taking shape. This was an active, committed, and participatory audience. Out of the camaraderies of the amateurs emerged more formal fraternities, the wireless clubs, which were organized all over America. One of the largest, formed in 1914 by the inventor

50

An early crystal detector, the inexpensive and reliable wireless detector most frequently used by "hams."

Hiram Percy Maxim, was the American Radio Relay League, which organized a national amateur network of stations across the country through which amateurs could relay messages to and for each other. Thus, by the mid-teens there existed in the United States a grass-roots, coast-to-coast communications network, as well as an incipient radio audience.

When the League was formed, *Popular Mechanics* proclaimed "the beginning of a new epoch in the interchange of information and the transmission of messages." The way these amateurs used the invention, trying to reach as many people and to be as inclusive as possible, was the opposite of the more closed, exclusive policies of the private companies and the navy. Through their activities, the amateurs raised the question: "Why restrict this invention to a few select corporate and military senders and receivers when so many everyday people could benefit from and enjoy this device?"

Amateur activity increased dramatically during the second decade of the century, and some of the more powerful stations transmitted voice and music. As early as 1909, the radio inventor Lee De Forest had begun using more sophisticated transmitting equipment to broadcast music and the human voice, and the amateurs' crystal detectors were capable of receiving such broadcasts. By 1914, De Forest was broadcasting voice and music fairly regularly from his station in Highbridge, New York, and other amateurs with similarly powerful equipment followed suit. By contrast, the wireless companies and the military stuck to sending the Morse code, and ignored this new use of radio.

Amateur stations were temporarily shut down during World War I, but when they returned to the air in 1919, the amateurs with access to transmitting tubes began broadcasting voice and music on a more regular basis. Other amateurs listened in and got their families and friends hooked on the hobby. It is important to emphasize that this way of using radio was completely at odds with how Marconi, the device's inventor, had envisioned its applications. He had seen radio as helping the military, shipping firms, and the press expedite the transmission of coded messages between specific senders and receivers. The broadcasting of voice and music was simply not part of his agenda: this was an innovation of the amateurs.

By 1920, there were 15 times as many amateur stations in America as all others combined. Yet the executives of the Radio Corporation of America (RCA), which was formed in 1919 to buy out the British-owned Marconi Company of America and to consolidate the radio industry in the United States, regarded its main business as the transmission of long-distance Morse code messages. By late 1920, however, with the amateurs leading a huge radio boom in the United States, RCA had to redefine its mission. The amateurs and their converts had constructed the beginnings of a broadcasting network and audience. They had embedded radio in a set of practices and meanings vastly different from those dominating the offices at RCA. Consequently, the radio trust had to reorient its manufacturing priorities, its corporate strategies, indeed, its entire way of thinking about the technology under its control.

By the 1930s, it was the major corporations, not the amateurs, who dominated America's airwaves. But a robust subculture of hams continued to transmit and to listen, especially with shortwave, and to tinker. One device they began tinkering with was the phonograph. By the early 1950s, this tinkering would revolutionize the recording industry in the United States.

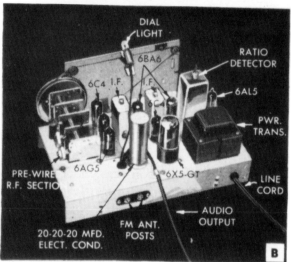

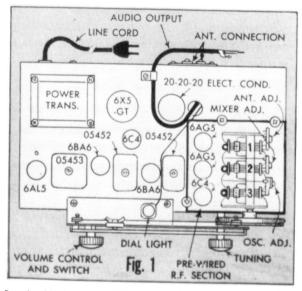

Popular Mechanics in 1950 provided instructions for building one's own FM receiver at home. The top photograph shows the completed tuner connected to an audio amplifier and a loudspeaker.

The Passion for Hi-Fi

"A new neurosis has been discovered," *Time* sarcastically exclaimed in January of 1957, "audiophilia, or the excessive passion for hi-fi sound and equipment." Sufferers were usually "middle-aged, male and intelligent, drawn largely from professions requiring highly conscientious performance." Six years earlier, *The New Yorker* had described the hi-fi craze as the fastest growing hobby in America. As early as 1952, the sales of hi-fi equipment to audiophiles had climbed to $70 million a year, and sales figures were still soaring. And this was before corporations began to manufacture and market sets for the general, non-tinkering public. By the mid-1950s, the phonograph industry which had, according to a September 1957 article in *Business Week*, "once looked down on hi-fi fans as mere fanatics," was scrambling to meet the new demand.

The hi-fi craze of the late 1940s and 1950s had been started by tinkerers dissatisfied with the sound quality available in commercially manufactured phonographs. So they began assembling their own "rigs" out of separate components, paying special attention to, and customizing, the wiring that connected the parts into a whole. The proper matching and balancing of components was critical to success. The goal was to reproduce in one's living room the way classical music sounded in a concert hall. The most sensitive human ear can hear sounds ranging from 20 to 20,000 cycles per second. Most old 78 rpm records could only play up to 7,500 cps, and AM radio could reach a maximum of 10,000, but usually broadcast at 5,000 cps. Audiophiles wanted to push beyond these restrictive ranges, which cut off the highs as well as the lows of most music.

This quest for fidelity gained impetus from several key developments just during and after World War II. The wartime shortage of shellac, the principal ingredient of records at that time, prompted research into other materials. The result was the introduction in 1946 of the vastly superior vinylite. Columbia Records used the material to introduce its new $33\frac{1}{3}$ rpm long-playing record in the spring of 1948. Using considerably finer grooves than the 78 rpm, the LP provided three to four times the playing time with considerably reduced surface noise, and with additional range and clarity. The LP could record up to 12,000 cps, twice the range of the shellac 78 record. In addition, the shift to magnetic tape in the late 1940s dramatically enriched the quality of recording. Yet most existing phonographs failed to do justice to the new LPs.

During the war, many servicemen and civilians were trained in the fundamentals of electronics in order to participate in the manufacture, installation, and operation of radar and other communication equipment. Those stationed in Europe, especially in England, became acquainted with the striking superiority of sound engineering abroad, and the significantly higher quality of music reproduction and phonograph equipment. After the war, when these men resumed civilian life, some brought imported audio components home, while others bought surplus amplifiers and other kinds of electronic gear from the government. Small electronics companies also began to improve amplifiers, speakers, and other components.

Armed with their recent training, soldering irons, miles of wires, and a host of experimental circuit designs, these veterans formed the initial core of the hi-fi enthusiasts who sparked the skyrocketing component parts trade of the late 1940s and early 1950s. The custom-built sets they assembled often provided twice the fidelity of reproduction that one could get from the most expensive commercial system, and for one-half to one-third the price. Magazines from *Popular Mechanics* to *The Saturday Review* began to run regular features on hi-fi construction, musical developments, and the intense technical debates that raged among hobbyists. In 1951, a new quarterly called *High Fidelity* began publication and in one year its circulation leapt from zero to 20,000.

The hobby's rate of growth was breathtaking, producing enormous sales for the small companies willing to cater to audiophiles by selling high quality components. By 1953, approximately one million Americans had invested in custom-built sets. Firms such as Fisher Radio Corporation and Altec-Lansing reported that sales had increased 20 times between 1947 and 1952. The quality of sound on these sets often produced instant converts: once

This 1950 cartoon from *The Saturday Review* pokes fun at the exasperated audiophile trying to assemble his own hi-fi set at home.

someone heard a record on a custom-built hi-fi, the listener had to have a set of his or her own. For those incapable of building their own sets, small firms such as Electronic Workshop would install a customized set. One repeatedly noted characteristic of the audiophile was that he was never satisfied; he was constantly striving for greater fidelity, and spent endless hours and hundreds of dollars a year trying to approximate perfection. He was also completely disdainful of corporate America's audio offerings.

As in the case of the amateur radio operators, there were barriers to entry to this hi-fi fraternity. Technical knowledge separated outsiders from those in the know. So did language, as a whole new vocabulary containing words like woofer, tweeter, preamp, and equalizer made discussions among enthusiasts unintelligible to outsiders. Especially alienated were women, who were not just excluded from such technical activities, but who had to put up with obsessive, monomaniacal tinkering that filled living rooms with boxes and wires and covered rugs with gobs of solder. Women began publishing articles with titles like "The High Fidelity Wife, or a Fate Worse than Deaf" and "I Was a Hi-Fi Widow."

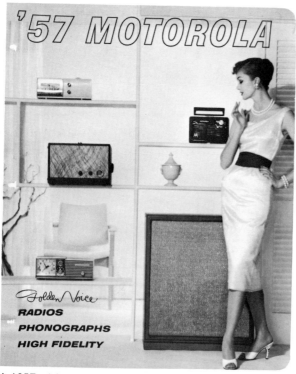

'57 MOTOROLA

Golden Voice

RADIOS
PHONOGRAPHS
HIGH FIDELITY

A 1957 ad for Motorola radios and record players, included a factory-assembled hi-fi set. Manufacturers sought to cash in on the hi-fi craze by marketing equipment to those who could not assemble their own, but who wanted the audio fidelity of a hi-fi set. Women were an important market for such sets.

While some hi-fi enthusiasts spent more time tinkering with their rigs than listening to the music they produced, what bound these men together was an obsession with musical authenticity. The hi-fi craze defined musical appreciation as a masculine enterprise best cultivated by those with specialized, often technical knowledge who could truly discern good music, not just with their hearts, but with their heads. Hi-fi enthusiasm allowed men to retreat from the wartime application of technology to cultural destruction, to the use of their electronic expertise in the service of beauty and cultural preservation. Through this hobby, so firmly embedded as it was in the masculine province of electronics, men could indulge in the pleasures of music while warding off suggestions of effeminacy. With charts, circuit diagrams, and potentiometers, enthusiasts could quantify music, move it from the emotional to the rational realm and make its appreciation objective and mental, not subjective and physical.

The connections such devotées made between the feminine and the inauthentic and, therefore, the inferior, became clear when the more entrenched manufacturers such as Magnavox and RCA sought to sell preassembled hi-fis to the general public and, particularly, to women. Entering the market belatedly in 1953 and 1954, these companies marketed their hi-fis in finished cabinets that matched existing furniture designs such as French Provincial or Colonial. Audiophiles denounced the new equipment as overpriced, inferior, too dedicated to appearance, and not even meriting the label hi-fi because they didn't come close to the rigid specifications of enthusiasts.

Another characteristic many of these enthusiasts shared was a deep aversion to the other new electronic invention sweeping the United States, the television. Their devotion to musical authenticity, and their antipathy to the passive, physically idle consumption of popular culture, made many of these audiophiles the first dedicated listeners to FM radio. The quest for fidelity, in other words, was not only a technical quest driving the improvements in hi-fi equipment and then in FM transmitting and receiving, but also a cultural and political quest for an alternative medium marked by fidelity to musical creativity and cultural authenticity. The quest for fidelity meant the reduction of noise, not just from static, but also from the hucksterism of America's consumer culture. This mindset, which was adopted and reshaped by the next generation of rebellious young men, helped spawn a new group of audio outlaws, the underground FM programmers of the late 1960s and 1970s.

The FM Revolution

From the earliest beginnings of its technical, business and regulatory history, FM was an industry outcast, an anti-establishment technology marginalized by vested corporate interests. Invented by Edwin Howard Armstrong in the early 1930s, FM was immediately perceived by David Sarnoff, the head of RCA, as a major threat to the already established AM industry. Sarnoff reacted by

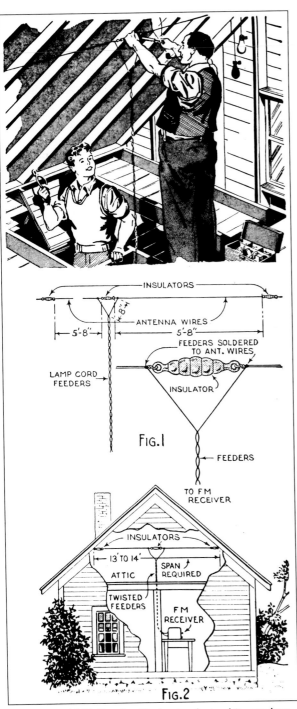

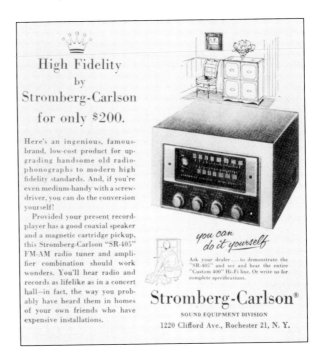

As early as 1943, *Popular Mechanics* showed its readers how to install an indoor FM antenna for maximum reception. *Top right:* An ad in *The Atlantic* from 1954 addressed the audiophile seeking to improve the fidelity of his existing phonograph and radio.

doing all he could to try to thwart the invention. He blocked financial support for experimentation, and he worked behind the scenes at the FCC to block allocation of spectrum for FM use. It is not surprising, then, that FM's renaissance would be pioneered by those very much outside of—even at odds with—the media culture those corporations had created.

Despite efforts to suppress his invention, Armstrong had by the early 1940s developed a small FM network in the Northeast and a small group of fans had acquired FM receivers. The FCC's decision in 1945 to reallocate FM's slot on the spectrum made those sets obsolete, and with FM's prospects seeming so uncertain, the number of stations actually declined in the early 1950s. Beginning in 1958, however, FM began to experience a resurgence. The number of stations began to increase, and so did the audience. The AM spectrum had become so crowded, especially in major cities, that by the late 1950s there were few or no slots left. The only way to start a new station was to use FM. Hi-fi enthusiasts began to tinker with FM, and others bought the newly available sets, especially imported ones. Between 1960 and 1966, the annual sales of FM radio receivers increased more than fivefold, and by 1967 over one-third of all radio sets sold were equipped with FM reception. In 1960, there were approximately 6.5 million households with FM; by 1966, that number had soared to 40 million.

These early listeners to FM stations were usually more educated than the average American, and tended to have "high culture" tastes, preferring FM's music, intellectual fare, and lack of commercialism to the usual AM programming. The households that accounted for the bulk of FM listening were also the ones that watched the least amount of TV and, in fact, turned on FM rather than TV during the prime time evening hours. FM audiences were concentrated in major metropolitan areas like New York, Chicago, Los Angeles, Washington, and Boston, and in the 1950s and early 1960s urban FM stations catered to their listeners' devotion to classical music. By the mid-1960s, however, 61 percent of FM stations played "middle of the road" music, which ranged from

Frank Sinatra and Mantovani to Dave Brubeck.

The immediate catalyst for the FM explosion in the late 1960s came from the FCC. Since the late 1940s, most of the FM outlets owned by AM stations had simply broadcast exactly the same programming as their AM parent. But by the early 1960s, FCC Commissioners Robert E. Lee and Kenneth Cox argued that frequencies had become so scarce that in the face of increasing demand, duplication was "a luxury we can't afford." In May 1964, the FCC issued its nonduplication ruling, which was to take effect in January 1967. In cities with populations of more than 100,000, radio stations with both AM and FM could not duplicate more than 50 percent of their programming on both bands simultaneously. This ruling helped promote much more enterprising exploitation of the medium: between 1964 and 1967, 500 new commercial FM stations, and 60 educational stations took to the air.

A handful of figures suffice to convey the mag-

By 1960, *Popular Mechanics* was promoting kits that allowed enthusiasts to produce a complete "stereophonic music system." Here, the magazine showed how such a system could provide soothing background music for the harried executive.

nitude of the FM revolution. In 1964, total net FM revenues were $19.7 million. Ten years later, that figure had increased thirteenfold to $248.2 million. In 1962, according to the FCC, there were 983 commercial FM stations on the air; in 1972, their number stood at 2,328. Four years later, there were nearly 3,700 FM stations on the air. By 1972, in cities such as Chicago and Boston, it was estimated that 95 percent of households had FM sets. A few years later, that figure held for much of the country.

While technical refinements, overcrowding in the AM band, and regulatory changes were obviously critical factors in the FM explosion, it was also the emergence of a profoundly anti-commercial, anti-corporate ethos in the 1960s that caused FM to flower. This ethos was marked especially by a contempt for what had come to be called "mass culture": a disdain for the "vast wasteland" of television and for the formulaic, overly commercialized offerings of radio. It also represented a scorn, first, on the part of older intellectuals and, later, on the part of the counterculture, for the predictability and mindlessness of mainstream popular music.

The rise of 1960s youth culture especially transformed FM's content and appeal. Bound together by rock and folk music, contemptuous of the commercialization that seemed to infuse and debase every aspect of American culture, and hostile to bourgeois values and the profit motive, members of that loose yet cohesive group known as the "counterculture" were revolutionizing almost every aspect of American culture, from its popular music to its language and clothing.

Particularly hateful to these young people was what they saw as the lockstep conformity of American life that made everything from work to popular music joyless, unspontaneous, and false. They wanted something different: they wanted their lives to be less programmed, less predictable. The music they were listening to, which was not broadcast on AM, gave expression to their critique of mainstream culture. At this time, AM radio was characterized by incessant commercials, songs lasting no longer than three minutes, and repeated promotional jingles. It is no surprise that when some of these young people, primarily men, worked their way into FM radio stations, they deliberately used their positions to challenge every aspect of what people heard and how they heard it on the airwaves. That challenge led to the proliferation of "underground" or "progressive" rock stations around the country.

Some of the earliest of these stations, which went on the air between 1967 and 1969, were KMPX in San Francisco, KPPC in Pasadena, KMET in Los Angeles, WOR and WNEW in New York, and WBCN in Boston. The rebellious young men manning these underground stations differed somewhat from the amateurs and hi-fi audiophiles. They were less interested in technical tinkering, in getting inside the "black box" of FM, than they were in using the invention for cultural tinkering to defy the establishment. When they started their own FM stations, they threw all the conventional industry rules and responses out the window. They eliminated advertising jingles, the repeated announcing of call letters, and the loud, insistent, firecracker delivery of AM disc jockeys. They repudiated conventional market research which sought to identify the "lowest common denominator" and thus reinforced the predictable repetition of the top 40 AM songs. College stations around the country, not surprisingly, pioneered and embraced the underground format.

Instead of being required to select songs only from a tight "play list" determined by a programming manager, disc jockeys on progressive rock stations were given wide latitude to play what they wanted. They also sought and responded to listener requests. They avoided most top 40 music and the playing of singles. Instead, a low-key, at times somnolent, male voice talked to the audience in what was called a "laid back" and intimate fashion in between long segments of music that included album cuts of rock, blues, folk, jazz, international and even, on occasion, classical music. Progressive FM stations especially delighted in playing the longer cuts of a song, some of them running as long as 12 or 20 minutes, for an audience that could hear such music nowhere else on the spectrum. In August 1969, *Broadcasting* labeled underground radio "the first really new programming idea in 10 years."

The majority of listeners to these stations were

An early "combination console" featuring AM/FM reception and a "multiple-play" phonograph with "Golden Voice" fidelity.

educated, affluent young men, and they were extremely loyal to such stations. Like their predecessors, the hi-fi enthusiasts, these men were dedicated to a musical cult of authenticity that emphasized the essential interconnections between composing, mastery of an instrument, and performance. The music they championed was usually complex, the lyrics metaphorical, political, or both, and spotlighted male virtuosos, especially on guitar or drums. Thus, while underground FM represented an explicit rejection of establishment notions of masculinity, it was also a deeply masculine enterprise focused on male performers, DJs, and listeners—all grappling with the crises surrounding traditional gender roles in the late 1960s and early 1970s. Progressive rock stations also specialized in information on the anti-war movement and general countercultural activities, rejecting the overly competitive and often destructive masculinity promoted in corporate and military circles.

Although underground radio represented only a tiny portion of FM stations, its impact on programming formats and content was enormous, precisely because it was so fresh, new, and compelling to listeners. In the 1970s, following this prolifera-

tion of stations and upheavals in program formats, the owners of FM stations saw an opportunity to make a profit. By October 1974, FM accounted for one-third of all radio listening, but only 14 per cent of all radio revenues. One reason that so much experimentation had been possible with FM was precisely that advertisers exerted very little influence over the medium. Prejudiced by the notion that FM listening was the province of "eggheads and hi-fi buffs," advertisers had eschewed FM until the early 1970s. But advertisers and owners of FM stations recognized that in spite of considerable alienation, American youth nonetheless constituted a big market, and as a result more and more stations converted to some type of rock format.

To appeal to the younger market, the ABC-FM network developed a hybrid format with the predictability of the AM format as far as music was concerned, but the underground style of announcing. In 1971, CBS-FM followed suit, co-opting some of the stylistic innovations of the underground while purging it of left-wing politics and too much musical heterogeneity. Such network initiatives exploited some of FM's iconoclasm in order to turn the anticorporate ethos to the industry's advantage.

In November 1974, *Broadcasting* featured an article entitled "FM Rockers are Taming Their Free Formats." The article noted that many progressive stations were adopting tighter playlists and starting to rely on market research. Albums out of the mainstream, once the mainstay of early FM, were now no longer given a chance at many stations. The playlist was agreed upon by a committee or determined by the program manager, as it had been in AM during the 1960s. Accompanying this trend toward homogenization was the adoption by different stations of a very particular, tightly circumscribed format: oldies, soft rock, album-oriented rock, or country and western, with very little, if any, overlap.

By the late 1970s, the assembly-line techniques that the early FM outlaws had deplored were now informing much of FM programming. As *Advertising Age* noted in May 1978, "The day of the disc jockey who controls his individual program is quickly becoming a dinosaur." As had been the case with the amateur operators and the hi-fi audiophiles, the defiance of early FM enthusiasts invigorated an entrenched and complacent industry; but this defiance was quickly domesticated in the quest for massive audiences and profits.

The tinkering of these audio outlaws set the stage for radio broadcasting in the 1920s, revolutionized the phonograph and recording industries in the 1950s, and pioneered the use of a whole new frequency band, FM, in the late 1960s and 1970s. All three groups of enthusiasts were outsiders who regarded the corporate uses of audio technology as unimaginative, technically backward, and culturally stunted. Each group, in its own mode, challenged the way the profit motive had circumscribed the exploitation of, and access to, audio technology. Each group was also testing the boundaries of society's definitions of masculinity and femininity, embracing the former through technical mastery, and exploring the latter by using their machines to achieve deeper emotional and interpersonal satisfactions. The "ham" operators still constitute a robust subgroup that exchanges messages around the world, proudly circumventing more established communications systems, while the more defiant technical outlaws have adopted the computer as their vehicle for fraternal rebellion.

Oppositional reactions against the dominant culture by technological enthusiasts have burst forward at various moments during our history. They represent serious, often passionately held views about what culture should be, and questions about the extent to which the demands of the marketplace should shape cultural practices and products. They also represent the vision of subcultural groups of men with often utopian ideas about how machines can promote a sense of community and reproduce cultural excellence. But one of capitalism's greatest strengths is its ability to incorporate the voices and styles of the opposition into a larger framework, and to adapt such opposition to its own ends. The cultural benefits are, of course, that mainstream culture does change, is enriched, and does, at moments of technological uncertainty and cultural upheaval, provide brief periods when diversity can really flower.

Some of the most revolutionary music of the 1960s and 1970s was first heard on FM stations.

At 4:35 in the morning the signal was given, and in an instant I am in the air, my engine making 1,200 revolutions, almost its highest speed, in order that I may get quickly over the telegraph wires along the edge of the cliff. As soon as I am over the cliff I reduce my speed. There now is no need to force my engine. I begin my flight, steady and sure, toward the coast of England. I have no apprehensions, no sensations. . . .

The moment is supreme, yet I surprise myself by feeling no exultation.

Below me is the sea, its surface disturbed by the wind, which now is freshening. The motion of the waves beneath me is not pleasant. I drive on. . . .

Now, indeed, I am in difficulties, for the wind here by the cliffs is much stronger and my speed is reduced as I fight against it, yet my beautiful aeroplane responds. Still steadily I fly westward, hoping to cross the harbor and reach Shakespeare cliffs. Again the wind blows. I see the opening in the cliffs. . . .

Avoiding the red buildings on my right, I attempt a landing, but the wind catches me and whirls me around two or three times. At once I stop the motor. Instantly my machine falls, straight upon the land, from a height of 65 ft., in two or three seconds, and my flight is safely done.

Louis Bleriot
Quoted in *Popular Mechanics*, September 1909

NINETY YEARS

OF

POPULAR

MECHANICS

MARY L. SEELHORST

Editor, *Popular Mechanics:*

I've been fascinated by aviation ever since I read
your September 1909 issue (which I still have). It
talked enthusiastically about such momentous events
as Louis Bleriot's crossing of the English Channel and
Glenn Curtiss' record-setting endurance flight of 52
minutes, 30 seconds.

Not everyone was as enthusiastic as you in those
days. A history of the United States published in
1911 had this to say about the future of aviation: "It
is hardly probable that the art will ever be of much
practical importance to the commercial world."

The continuing excellence of your aviation articles has certainly helped to encourage enthusiasm
for one of man's greatest achievements.

Clarence Jefferson,
Linden, New Jersey
December 1984

Perhaps no single source can better inform our understanding of American enthusiasm for technology than *Popular Mechanics Magazine*. Enthusiasts of all types have dog-eared its pages since the beginning of this most technological of centuries. Immensely popular among men and a staple of barbershop libraries, *Popular Mechanics* is today emblematic of America's fascination with technology. Written by and for enthusiasts, this magazine has reflected and reinforced the country's love of technology since 1902.

From inauspicious beginnings in 1902, the magazine matured quickly. Within a few years *Popular Mechanics* had found a nationwide audience and begun to reflect a society—and indeed a world—changed by changing technology. "It is the Twentieth Century paper," claimed a first-year issue. "The regular reader of *Popular Mechanics* need never be at a loss for entertaining things to talk about." Henry Haven Windsor, the founder and first editor of *Popular Mechanics* (in later years also referred to as *PM*), recognized that in addition to its potential to change the way people lived and worked, technology could be a source of entertainment. "Yankee ingenuity" notwithstanding, most technical publications available at the turn of the century were trade journals for men working in a specific mechanical field; few magazines attempted to promote technical knowledge and skills to a general audience. Shortly before his death in 1924, Windsor stated:

I started *Popular Mechanics Magazine* twenty-three years ago with the idea primarily in mind that it should be a magazine of educa-

Henry Haven Windsor had high hopes for the little weekly that he founded in 1902. *Popular Mechanics* would explain "the way the world works" in "plain, simple language," incorporating photographs and diagrams to help readers understand complex topics.

tion, and that while up to that time there was practically nothing of a technical nature set forth in plain, simple language, in other words, "written so you could understand it," I hoped to accomplish that very thing.

"Written So You Can Understand It" became the magazine's slogan, optimism and enthusiasm its hallmarks. By 1910 the magazine was a phenomenon, circulation was a quarter million a month, and Windsor was a millionaire. Flattery by imitation followed. Its predecessor in the popular coverage of science and technology, *Popular Science,* added do-it-yourself topics in 1915. Several other magazines were later founded in the same genre, including *Mechanix Illustrated* in 1928; *Everyday Mechanics* (later *Science & Mechanics*) in 1930; and *Mechanics and Handicraft* in 1933.

While today dozens of popular technical titles crowd the magazine racks at supermarkets, most

are specialty magazines focusing on a single field of interest such as computers, photography, woodworking, or automobiles. Yet *Popular Mechanics* remains a classic of the genre; a generalist among specialists in a highly competitive field of publishing. To this day it has maintained a truly popular mix of information, instruction, and entertainment, building an ongoing dialogue with a large community of enthusiastic and involved readers. Most analyses of *Popular Mechanics*, both academic and popular, have focused on relatively narrow aspects of its wide-ranging technological lore: predictions of the future, do-it-yourself projects, or wacky widgets of little practical value. Although all of these things have been important to its success, none of them can adequately characterize its scope, or in itself explain the magazine's continuing appeal to readers.

Non-readers may be surprised at the breadth of the magazine's offering. Anyone who has seen even a single issue knows that its contributors get behind the walls, under the hood, and inside the workings of inventions, machines, and technological systems—both domestic and foreign. But to Windsor, "mechanics" meant not only the nuts and bolts of machinery, but also "the way the world works." As a result, readers can find articles such as "Radium, the Metal of Mystery" in 1904; "At the Bottom of the World with Byrd" in 1930; "The Air Around Us: How It Is Changing" in 1964; and "New Views of Mars" in 1989. As new inventions and discoveries expanded the scope of man-made things and the term "mechanics" gave way to "technology," *PM* kept

POPULAR MECHANICS

An Illustrated Weekly Review of the Mechanical Press of the World

VOL. I. NO. I.   CHICAGO, JANUARY 11, 1902.   PRICE 5 CENTS.

Illustrations of new and unusual things, like this interior of the British Navy's first submarine, appeared on *Popular Mechanics* covers from the first issue. *PM*'s distinctive covers were later widely imitated in the popular technical press.

pace. Articles like "Trouble-shooting Ford's Microcomputer Control Unit" (1982) joined titles in the more traditional realm of practical culture, while in 1991 "The Lessons of the Black Box War" explained how electronic devices helped determine the outcome of the war in the Persian Gulf, once again changing "the way the world works."

Since *Popular Mechanics* was a picture magazine from the beginning, it is not surprising that the most obvious signals of its purpose over the course of 90 years are visual. Particularly telling are *PM*'s signature covers, which present images of changing technology in a graphic style that reflects current trends in magazine design. There, images, style, and words combine to convey the magazine's consistent fascination with technological change. But in order to be successful, a popular magazine that predicts and marks the advance of technology must also be relevant to its readers' lives. The front half of *Popular Mechanics*, the traditional location of scientific and technological news, is a litany of notices on what is new and different; the back half of the magazine has always been the location of practical, how-to information of all types. This is the line *Popular Mechanics* has negotiated with such skill for the last 90 years—the line between its readers' dreams and realities. Indeed, *Popular Mechanics* is as much about its readers as it is about mechanics.

Even a cursory reading of the magazine reveals that *PM* readers share an avid interest in self-improvement, whether by making things, buying things, or simply learning about the latest developments in science and technology. This message comes through clearly in the headlines of articles: "What Television Offers You" (1928); "How will automation affect your job?" (1962); or "Soup Up Your Home Computer" (1985). During almost a century's worth of radical technological change, *Popular Mechanics* has continually addressed the questions of "What's new and how does it work?," "Can I do it myself?," "How will it affect me and my family?," "How might it change our world?," and has found an audience eager to know the answers.

Of the current readers, 86.5 percent are male; earlier statistics are more elusive. While *Popular Mechanics* has always been considered a men's magazine, the readers it courts today—and has courted throughout most of its history—are middle-class men with homes and families. From time to time, regular columns addressing the interests of women or children have surfaced and then disappeared. But whether they picked the magazine up or not, wives and children of avid readers probably found their lives influenced by its presence in the home. Family-oriented topics are abundant throughout most of the magazine's history, lending credibility to a claim in a *Popular Mechanics* advertising rate flyer of 1952 that 78 percent of *PM*'s readers were married, and "They use our pages to buy for their wives, their children, their homes."

For better or worse, the magazine has reflected and reinforced our society's gender and class biases concerning technology as much as it has reflected and rein-forced our enthusiasm for technology. And whatever else they are, the readers of *PM* are certainly technological enthusiasts. Technological possibility has motivated inventors, inspired hobbyists, encouraged consumers, and engaged children. While we may never be able to gauge the extent of the magazine's influence, it is clear that the changing but ever present face of American technological enthusiasm has been captured in the pages of *Popular Mechanics*.

"Hitherto Unprecedented"

A series of radical inventions and scientific discoveries startled the world around the turn of the 20th century. Among them were the discovery of X-rays in 1895 and radium in 1898, the invention of wireless communication in 1895, and the first modest success of the heavier-than-air flying machine in 1903. In the midst of this rapid-fire inventing, Henry Haven Windsor established and refined his new magazine like an experimenter developing a new apparatus. From the beginning Windsor had a good hunch of what his readers wanted. Although he made some minor adjustments during the first few years of publication, there were no fundamental changes in what had proven to be a successful mix of technological news, practical tips, and hands-on projects—a combination that continues to work to this day. While changes in technology, labor, modes of production, and available goods were certainly factors that affected the content of the magazine, no analysis could be complete without recognition of Windsor's tremendous influence on the tone and direction of

PM's cover art tells much about the content of the magazine. While the technologies depicted have changed dramatically, the cover art continues to express a high degree of enthusiasm and interest. New ideas, domestic comforts, futuristic proposals, and military might are popular topics. *Clockwise from top left:* June 1920; October 1953; May 1974; April 1991.

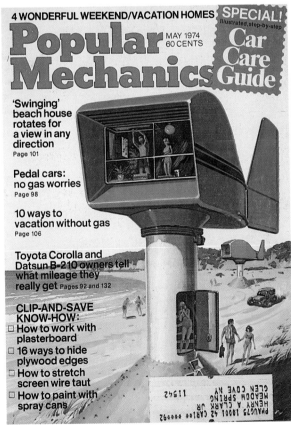

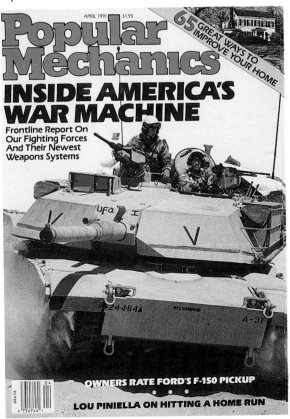

FEED ALARM FOR SLEEPY MILLERS.

The accompanying wrinkle comes anonymously from Seneca Falls, N. Y.: It requires no explanation. The flow of the feed in the feed hopper, by gravity, causes a dingus, provided with an eccentric, to revolve. When the feed ceases, the dingus

stops revolving, the eccentric slips and a ten-pound weight drops on the head of the sleeping miller. This will cause the miller to awake. If it does not, the weight should be increased. The dingus can be made of any metallic substance. The miller should be of combustible material so that he can be fired.

Humorous notices like this 1904 contribution often appeared in early issues.

Popular Mechanics during his 22-year editorship.

Unabashed awe pervaded the magazine's early coverage of the important discoveries and applications it later described as "wonders of the modern world." Given that the practicality of these inventions was still in question, it is not surprising to find that amusing novelties—like the "Feed Alarm For Sleepy Millers" published in 1904—often received equal billing with significant inventions. In what could be termed the magazine's "era of superlatives," a preponderance of dramatic headlines and short articles tended to convey amazement rather than in-depth understanding. Covers often pictured the biggest, smallest, and oddest of everything imaginable from ice blocks to submarines: "Queerest Water Works in the World" (1902); "America's Newest and Deadliest Submarine" (1903); "Largest Automobile in the World" (1904).

Major feats of civil engineering, such as the Panama Canal and Egypt's early Assouan (Aswan) Dam, came in for a large share of attention, but no locale received as much attention as the city of Chicago—the headquarters of *Popular Mechanics*. Windsor mined his own back yard for many cover stories in the early years, fortunate that Chicago was one of the most dynamic cities in the nation when it came to architecture and engineering. Nine of the 1902 weekly covers featured the drama of that city's new man-made landscape with headlines such as "Oddest of Chicago's Odd Bridges," and "Great Steel Structures on State St., Chicago." Windsor may have been quite deliberately appealing to what was then a largely midwestern readership in order to build circulation.

But the amount of attention paid to Chicago also suggests that many of the enthusiasms and opinions published in *Popular Mechanics* in the early years were Windsor's own. Although the gen-

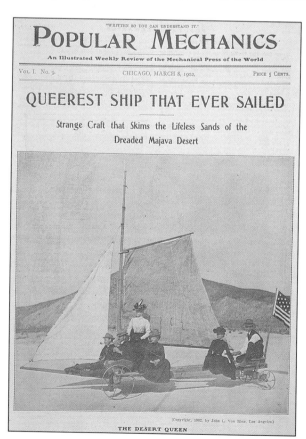

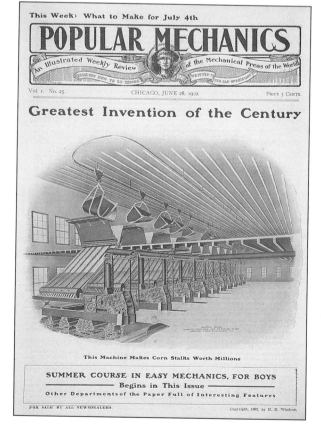

POPULAR MECHANICS
An Illustrated Weekly Review of the Mechanical Press of the World
Tells you how to do things · *Written so you can understand*

Vol. 3. No. 4.　　CHICAGO, JANUARY 24, 1903.　　PRICE 5 CENTS.

Royalty Opens the Great Assouan Dam

The Khedive of Egypt, the Duke of Connaught and Others of the Nobility Traversing the Dam on Trolleys

IN THIS ISSUE:

Pictorial Description of the Nile Dams.
Opening of the Assouan and Assiout Dams.
Traction Engine That Walks.
Carbon Paper Making, a Carefully Guarded Secret.
Tells Where the Trains Arrive.
Electric Railway Series—Part III.
Four Wheel Drive for Autos.
Handles Grain Automatically.
Concrete Bridges Gain in Favor.
Are Diamonds Imported From Other Worlds?
How to Dispose of Soot.
Purification of River Water.
Crude Dairies in Porto Rico.
Easy Electrical Experiments:—Designs of a Small Dynamo.

Novel Brick Conveyor.
How to Build a Sheet Metal Boat—Part II.
Hurls Water to the Sky.
Tiny Motor of Great Power.
Withstood Electric Current of Enormous Force
Handsome British Turbine Yacht.
New York's Subway in the Air.
Flashlight Work.
Shop Notes:—Simple Hand Bending Tools—Smelting By Electricity—Boiler Tube Cleaner—Telescoping Air Jack—The Injector Condensor—Jumping Up Iron for Welding—Simple Drawing Board Attachment—To Remove Paint by Sand Blast—What the Fly Wheel is For—Simple Motor Starting Device for Autos.

POPULAR MECHANICS
An Illustrated Weekly Review of the Mechanical Press of the World
Tells you how to do things · *Written so you can understand*

Vol. 2. No. 13.　　CHICAGO, SEPTEMBER 27, 1902.　　PRICE 5 CENTS.

Oddest of Chicago's Odd Bridges

Queer Modern Structure Built on the Principle of the Ancient English Drawbridge

IN THIS ISSUE:

Modern Bridge from Ancient Design.
Tunnels Under Cities Great Dangers in Time of War.
Locomotives of a Century (continued).
Great Demand for Young Men in Engineering Work.
French Electrical Instrument to Revive Drowned Persons.
Finest Brown Paint Made from Mummies.
How to Make a Mariner's Compass.

Shop Notes:—Device for Washing Overalls—Stripping Silver from Plated Articles—New Rotary Synchronizer—How to Make a Sparking Coil—Another Solder for Aluminum—Swaying of Belts, and Cure—Nitric Acid Made from Air—Dinner Pail Heater—Solid Foot for Ladders.
Bears Predict a Cold Winter.
Radical Change for Long Electric Roads.
Hollow Axles for Automobiles.

Popular Mechanics can be bought of any newsdealer in the United States.

Facing page and above: Some first-year cover headlines accurately described the novel constructions depicted; others embellished reality. A machine that made paper pulp from corn stalks was hailed as the "Greatest Invention of the Century"—a daring claim, even at the beginning of the century.

eral tone of the magazine in its first four years has been compared to that of *Ripley's Believe It or Not*, *PM*'s editorial opinions were not entirely celebratory or uncritical. For a 1904 article about a new Standard Oil Company scheme, "Rockefeller Will Pipe Oil Across America," an unnamed writer painted an unflattering picture of the largest oil refinery in the world:

> Smoke that rolls from a thousand stacks and, commingling into one dense, massive cloud, is wafted out over the lake, enshrouding water and commerce in its gloom: . . . sedulous, grimy streets as black as night, a city of perpetual exertion . . . .

But it was the how-to aspect of the magazine that gave the publication its unique status in the ranks of the popular press. "Shop Notes," the long-lived column started in the first issue, at first consisted primarily of general hints and tips useful to men working in mechanical trades: "Tool Rack for Engine Room," "How to Make a Steam Blower," and "Tinning Cast Iron." Windsor's experience as editor of such trade publications as *Brick* and *Street Railway Review* may have influenced his early focus on trade, rather than home tips. But soon, many of the tips published in "Shop Notes" were coming from readers. "When you happen on something good in your own experience let us give others the benefit of your skill," read a 1904 notice. Slowly the column reoriented toward tips for the home handyman.

During this period, Windsor was successfully exploring new markets, especially the country's large rural population distributed on farms and in small towns. In 1906 he ran a notice in one of his trade publications, *R.F.D. News*, offering rural postmen a new mail wagon if they sent him 100 paid subscriptions to *Popular Mechanics*. By 1909 about half the content of "Shop Notes" was home or farm related, with titles like "How to Make a Brooder" or "Interior Clothes Line"; within a decade, trade tips had almost disappeared. Advertising, confined to a separate section, addressed *PM*'s large new audience through mail-order ads for everything from tonics to trade schools to engines.

Windsor's desire to produce an educational magazine went beyond the inclusion of "Shop

Two Boys Build an Automobile

The accompanying engravings show the completed work of twin boys, Wilford and Winfurd Goddard, 15 years of age. The boys started out with no other material than what they could collect around their own home. No suggestions were received by them and they designed and completed the work of building an automobile with the exception of the gasoline engine. This engine they purchased from their earnings. The automobile is about 8 ft. long, with a 40-in. tread. The driving arrangement from the engine to the rear axle is connected to a cone clutch, which in turn is connected to a chain drive. The wheels were made from large carriage wheels cut down to the proper size and fitted with 28-in. rims. The tires are standard bicycle tires with an extra cover. On a trial trip it carried four boys 6

The Twins and Their Machine

miles, up and down hills and over sandy roads, at a speed of about 10 miles an hour.

◆ ◆ ◆

A scientist has calculated that the eyelids of the average man open and shut no fewer than 4,000,000 times in the course of a single year of his existence.

Side View: Seat Removed to Show Construction

Assisting the mechanical efforts of amateurs, whether boys or men, soon became a chief aim of the how-to portion of the magazine. Their sophisticated machines were also featured, as in this 1908 editorial.

Notes" and trade school ads, however. His first issue initiated the series "Easy Electrical Experiments for Boys," and subsequent articles frequently celebrated boys' achievements under such titles as "How a Boy Made a Set of Wireless Telegraph Instruments" in 1902, and "Two Boys Build an Automobile" in 1908. Indeed,

Windsor promoted the magazine as a practical and wholesome activity for boys. "Why not divert the young man's thoughts in practical and useful channels?" asked a 1903 advertisement. "Most boys would rather read *Popular Mechanics* than any story paper."

As with "Shop Notes," however, the columns featuring projects for

boys were soon reoriented to appeal to a broader audience. Within a year, the phrase "for Boys" was dropped from "Easy Electrical Experiments." By 1904, a number of simple projects were grouped under the heading "Mechanics for Young America," and by 1907 the section was renamed "Amateur Mechanics," a designation it was to keep for many years. Although boys were still pictured building many of the magazine's simpler projects, the changing headings reflect what must have become increasingly apparent to Windsor: a person could be an amateur at any age. An experienced boiler operator could be a novice at woodworking; a successful farmer might not know much about electricity; a professional office worker might enjoy building a glider. For several years Windsor continued the process of focusing his magazine to appeal to a broad, popular audience.

"The world has become mechanical."

The tone of amazement typical of *PM's* early years abated somewhat as the circulation of the magazine grew from about 70,000 in 1906 to over 200,000 in 1909. At first Windsor and his small staff culled articles from other publications with a mechanical bent, but increasingly he commissioned stories exclusively for *PM*. Feature articles and do-it-yourself plans became longer and more detailed, the advertising section grew fatter, and the magazine's staff expanded.

By that time too, many late 19th- and early 20th-century inventions were being put to use in a variety of ways. Between 1909 and the end of World War I, the news portion of the magazine tended to focus on practical applications of various technologies, often speculating about their potential in the future. In 1912 the editors polled nearly 1,000 scientists to select "The Seven Wonders of the Modern World": wireless, telephone, aeroplane, radium, antiseptics and antitoxins, spectrum analysis, and X-ray. An exuberant assessment compared them to the wonders of the ancient world:

> To the Ancients, a wonder had to be fashioned with the strong arm; its virtues were chiefly those of size and strength. The Modern Wonders find their inspiration in the service of human life—every human life—and their conception in minds, not in muscle.

One hotly debated technology was the flying machine. While the magazine and most of its readers expressed optimism about this invention, many people thought that the aeroplane would never be anything more than a plaything for the daring because it would be unable to carry enough weight or cover enough distance to be practical. But with a string of dramatic successes in the years around 1909, pilots became public heroes. Many inspired enthusiasm for aviation with their daredevil performances at popular aviation meets. While most people participated only vicariously, do-it-yourselfers could—and several did—build Alberto Santos-Dumont's famous Demoiselle monoplane from detailed plans published in *Popular Mechanics* in 1910. This celebrated Brazilian aviator had not patented his design, giving his plans "to the world in the interest of aeronautics."

Not only did *Popular Mechanics* report on technical developments worldwide, but its writers often claimed that modern technology would make people's lives better, and the world smaller.

Popular Mechanics covered the accomplishments of such pioneer aviators as Glenn Curtiss, Santos-Dumont, Orville and Wilbur Wright, and Louis Bleriot. But it was the latter's successful 1909 flight across the English Channel that made the world sit up and take notice, touching off a debate in *PM* over the use of aeroplanes in war. "Powerful Naval Fighting Machines Will Always Determine Destinies of Nations, Unmenaced by Aerial Craft" proclaimed the subtitle of a navy captain's article. In rebuttal, a pilot replied "The aeroplane is not simply the addition of a new device, it is more than this. It is the addition of a new field of operation." Within a few years, World War I would prove the pilot right.

The aeroplane was not the only "new field of operation" developing during the 'teens. *PM* in 1912 reported on the job of a load dis-

The Operator Waiting for a Call.

The Subscriber Calls; a Small Lamp at Switchboard Lights; Operator Plugs Calling Number, Extinguishing Light.

Operator Plugs Called Number; Bell Rings and Switchboard Lamp Lights and Burns Until Subscriber Answers.

Subscribers in Communication With Each Other.

Subscribers Hang Up Receivers; Lamps at Switchboard Light and Burn Until Operator Withdraws Plugs

WHAT HAPPENS WHEN YOU CALL "CENTRAL"

Just what occurs in the telephone company's establishment when one calls for and obtains a telephone connection is to most of us a mystery which the illustrations on this and the opposite page go far to clear up. Incidentally, a study of the complicated process of "getting a number" may serve to make one less impatient at delay or occasional mistakes on the operator's part.

The illustrations on this page show what occurs when the call is for a subscriber whose phone is connected with the same exchange as that of the "party" calling. On the opposite page are shown the operations necessary to establish communication between subscribers connected with different exchanges.

❡The hairlike tips of the heads of grain, often called "beard," and like formations on plants and trees, act as outlets for the earth's electricity, causing them to be always surrounded by a static charge, according to a Finnish scientist. This, he claims, has a beneficial influence on the growth of the plant, and explains why there are large quantities of ozone in pine forests.

As technological systems developed, people were able to use complex devices without understanding how they worked. Diagrams like this 1913 offering took some of the mystery out of everyday transactions.

patcher in the electrical power stations in "How The Modern Jove Juggles His Lightning"; reviewed the effectiveness of Cincinnati's new water filtration plant in 1913; and published illustrations explaining "What Happens When You Call Central." These and many other articles underscored Americans' increasing dependence on technology as inventions were transformed into operating systems that influenced ever-widening circles of commercial, political, and daily life. Conflicts among user groups became common, and *Popular Mechanics* often reported on laws and standards that attempted to regulate competing interests. Many of these governed safety.

While disseminating and celebrating technological advances, *PM* from the outset published articles on the risks attached to the application of these new technologies. Graphic depictions of aeroplane disasters, train wrecks, and automobile accidents appeared regularly, along with accounts of the threat posed by experiments with—and medical applications of—radioactive materials. Preventive measures were often proposed or reported: build metal rather than wooden train cars to reduce their vulnerability in accidents; improve road signs to warn motorists of dangerous curves; develop better shielding for X-ray machines and their operators.

There was even a proposal to abandon auto racing from no less a personage than Barney Oldfield, the most famous race car driver of the day. Though Oldfield continued to pilot race cars, his lengthy cover story of 1911 asked "Is the Game Worth the Candle?" In what was clearly an attempt to promote racing reforms, Oldfield described in detail some of the hundreds of deaths of drivers, mechanics, and fans that had occurred over auto racing's short history. While some fatalities were the result of crashes caused by mechanical failures, many more were easily preventable, having been caused by inadequate protection for spectators, blinding dust created by dirt tracks, and inexperienced but enthusiastic drivers. Though something of a racing promoter himself, Oldfield threw part of the blame on those promoters who profited by satisfying the "blood-hunger of the spectator."

Opinions were something the magazine promulgated in abundance up to its founder's death in 1924. Starting in 1909, with the magazine's popularity already well established, Windsor began a monthly editorial column soon titled "Comment and Review." While some columns celebrated mankind's progress in various technological fields, many tackled political, economic, and religious issues, offering views that might surprise today's readers because of their partisanship. Windsor, for instance, urged the repeal of blue laws (enforcing puritanical behavior) because he saw in them a dangerous union between church and state. He also advocated handgun control to reduce the number of violent deaths. Occasionally his editorial comments seemed at odds with the thrust of the magazine. Even as Dreadnoughts, torpedoes, and other naval ordinance competed for cover space during the intense international arms race preceding World War I, and the potential use of the aeroplane in war was enthusiastically debated in issue after issue, Windsor astutely observed:

The competition of nations in building more and larger ships than their fellows is approaching the day when the burden will chafe the neck of the people. In time of actual hostilities, the soldiers of Valley Forge heed not their bleeding, frozen feet; but in time of peace the element of consuming patriotism sleeps, and the shoe pinches.

As he predicted, his own patriotism came to the fore in 1917 with America's entry into World War I, and led to several passionate outbursts against the "barbaric Huns."

In the main, however, Windsor's stated opinions were confined to his monthly column. After his

On November 16, 1902, some 25,000 spectators lined Coney Island Boulevard in Brooklyn to watch gasoline, steam, and electric racing cars run against the clock over a one-mile course. This electric racer, built by automotive pioneer Andrew Riker, covered a mile in 63 seconds (57.1 m.p.h.). Journalists widely reported Riker's accomplishment as having established a world speed record for electric autos, and photographs of his tiny racer appeared everywhere, including the first issue of *Popular Mechanics*. *Left:* By 1911, auto racing had become a popular, though extremely dangerous, sport.

Above: When war is in the air, December issues of *Popular Mechanics* have often described the latest military toys. Ever-popular toy soldiers have fought "bloodless parlor-floor battles" for generations. *Left:* A Canadian in the trenches of France sent a letter of appreciation to the magazine in 1917, enclosing this photo. Many soldiers learned new mechanical skills, which they later brought back to civilian life. *Below:* Poison gas was used for the first time during World War I, requiring innovative defenses for civilians as well as soldiers.

death the "Comment and Review" feature was dropped, and subsequent opinions and predictions expressed in the magazine tended to be those of technological experts, famous achievers, business leaders, authors, or politicians who wrote for, or were quoted in, the magazine. *Popular Mechanics* continued to cover the technological news of the day and, for the most part, never promoted any particular political or technological cause over another. Although the general enthusiasm expressed in most of the articles could be construed as purveying opinion, the magazine often expressed the same kind of interest in directly competing technologies. Wartime, however has always been an exception. Then, in addition to showcasing nonclassified war technology, *PM* publicized government policy and generally promoted patriotism. The other exception has been the continual promotion of self-improvement for its readers by *Popular Mechanics*, a trend that intensified in the 1920s.

May   15 Cents

THIS ISSUE 435,000 COPIES

POPULAR MECHANICS MAGAZINE

WRITTEN SO YOU CAN UNDERSTAND IT

War-Zone Children Wear Gas Masks to School—*Page 691*

Popular Mechanics Offers No Premiums; Does Not Join in ... ... ys No Subscription Solicitors

N.S.E.

NOVEMBER   Super-Tone Radio—Page 829   25 CENTS

POPULAR MECHANICS
MAGAZINE
WRITTEN SO YOU CAN UNDERSTAND IT

REG'D. TRADE MARK, GREAT BRITAIN, No. 410426   REG. U.S. PAT. OFF.

War and the threat of war have long spurred technological innovation, but never with such far-reaching implications as in the 20th century. *PM* covered military-driven developments worldwide, as in this 1925 cover depicting a British proposal to launch and land aeroplanes from giant dirigibles.

"Easier Ways of Doing It"

When wartime restrictions were lifted in 1919, manufacturers applied technology to civilian life with unprecedented zeal, and Americans purchased the new and improved devices as never before. In contrast to the articles on major inventions and novel construction projects so prevalent in *PM* before the war, the 1920s pages were filled with news of incremental improvements to technologies once considered radical, but now becoming available to large numbers of middle-class people—items such as radio receivers, electrical appliances, and automobiles. Articles on buying, building, and repairing these devices became much more common. Amenities for the public, too, were important for towns hoping to attract a new kind of visitor—the auto tourist—and *Popular Mechanics* regularly reported on "Civic Features That Promote the Comfort and Enjoyment of Visitors and Residents." In public and private venues, the technology of cleanliness, convenience, and comfort could now be enjoyed by millions of Americans.

There is no better example of America's increasingly technological style of living in the 1920s and '30s than the growing electrification of homes and farms, a trend clearly encouraged in *Popular Mechanics*. Power plants, whose construction the magazine had so closely followed during the first two decades of the century, were finally paying off. A post–World War I suburban housing boom resulted in the design of thousands of completely electrified new homes promoted as models of efficiency and sanitation. Billing electricity as a "silent servant of mankind," *Popular Mechanics* published articles with such enthusiastic headlines as "Electricity to End Farm Drudgery" (1925), and "This Electrified Home Runs Itself" (1934).

Some drawbacks were acknowledged: a photo spread of the 1920s titled "Niagara's Beauty Menaced by Power Plants," asked "Will Commercial Development Such as This Destroy the Cataract's Beauty?" More immediately frightening to readers was the threat of electrocution, reported in such news items as "Electric Alarm Kills Man By Excessive Voltage" (1922). But readers could also learn about preventive measures in "It's easy to dodge Electric Shocks (if you know how)" (1927). Another ominous note came with the recognition that much of America's electrical power derived from non-

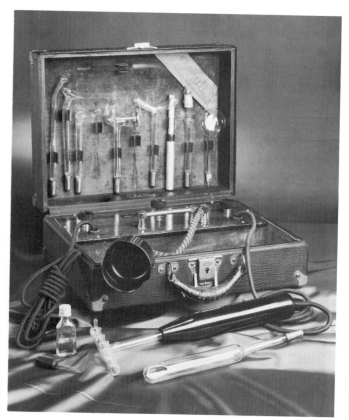

In the 19th century, many inventors and physicians believed in electro-therapeutics—the power of electricity to cure disease and invigorate the human body. Doctors concentrated electrical current, heat, and various types of rays on afflicted body parts. Several manufacturers marketed a "high-frequency apparatus" used to apply current to the body via specially shaped glass electrodes. In the early 20th century, widely advertised home versions were called "violet ray" machines, suggesting ultraviolet light, X-ray, and electrical power. Despite their popularity, government regulation and increasing consumer awareness gradually drove these devices off the market.

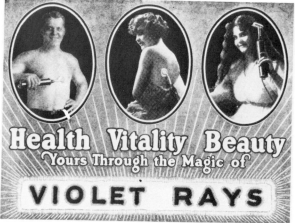

Health Vitality Beauty
Yours Through the Magic of
VIOLET RAYS

Most of us think of television as a post–World War II development of electronic technology. Yet several inventors successfully transmitted pictures in the mid-1920s by employing the mechanical scanning method. American inventor C. Francis Jenkins developed and constructed many parts of his system, including the precise circular glass prisms for his prismatic disc transmitter. Extensive press coverage of his work, such as this 1923 *PM* photospread, kept up public interest in this speculative technology. While electronic, rather than mechanical, television eventually became the industry standard, Jenkins had provided valuable research into optical transmission.

Movies in Home, Forecast of Inventor

The Heart of the New Radio Motion-Picture Receiving Apparatus

The Inventor, C. Francis Jenkins, Demonstrating the Radio Motion-Picture Camera; with This Device, Movies May Soon Be Enjoyed in the Home

A Contrast between His First Motion-Picture Projector, Invented by Mr. Jenkins in 1893, and the Essential Parts of the Radio Movie Machine, Which Is Rapidly Being Perfected by the Inventor

renewable resources, as noted by Edison Pettit of the Mount Wilson Observatory in a 1932 article, "Electricity from the Sun." "Sooner or later we shall have to go directly to the sun for power," wrote Pettit, "This problem of the direct conversion of sunlight into power will occupy more and more of our attention as time goes on, for eventually it must be solved." Despite an occasional alarm, electricity was generally hailed for its power to reduce the drudgery entailed in many tasks and its presumed potential to free mankind for higher pursuits and a better life.

But that better life depended on buying or making the devices that would transform flowing electrons into the "new chore boy." The regular two-page photo spreads initiated in April 1913 to feature convenience items for the home were now dominated by electric appliances like knife grinders, floor buffers, and dishwashers. Their titles, like the appliance advertisements of the era, emphasized ease and convenience: "New Ways to Avoid Home Drudgery" and "Time and Money-Saving Tools for Woman's Workshop in Home." Although the *Popular Me-*

chanics of this era undoubtedly fell into the consumer magazine category, today's consumers would miss one conspicuously absent piece of information—the name of the manufacturer! Nearly every issue carried some version of the following notice:

In accordance with the editorial policy of this magazine never to accept compensation in any form for what appears in our reading pages, and also to avoid all appearance of doing so, we are obliged to omit the name of the maker or the seller of any article described. This information, however, is kept on file and will be furnished free . . . .

Americans were enthralled with wireless communication from the moment Guglielmo Marconi introduced the new technology to America in 1899. Commercial stations, following the lead of amateurs, began to broadcast regular entertainment programming in 1920. Soon Americans were clamoring for the living-room boxes that could pluck far-away lectures and music from the air. The Radio Corporation of America quickly perceived profit in the fad, and soon marketed receivers designed for home use, such as this 1924 Radiola III-A. It was lashed to a pack horse and taken into the Rockies as part of *PM*'s 1925 "operative experiment, that in which a new instrument is tried out under novel conditions and in the face of difficulties possibly hitherto reckoned as prohibitive."

April • RADIO FOR PHONE USERS • N.S.L. 25 CENTS

POPULAR MECHANICS MAGAZINE
WRITTEN SO YOU CAN UNDERSTAND IT

SEE PAGE 547

The homeowner of the 1920s, however, was not just in the market for ready-to-use showroom machines. Many of the new conveniences were expensive, and *PM* published dozens of plans for homemade versions. "The domestic handy man will, if he is ambitious, sooner or later try to build a washing machine" claimed a 1924 plan for an electric washer housed in a galvanized garbage can. Many readers sent in tips on how to combine one or two electrical appliances with other parts to make a third—such as the 1922 "Hair Drier Made from Electric Toaster and Vacuum Cleaner"; or the "Small Electric Stove Improvised from Old Tin Can and Flatiron" contributed by a New York woman in 1925. Electricity also reshaped the home workshop, and *Popular Mechanics*'s advertising section filled with pitches for various electric hand tools.

Unprecedented numbers of urban dwellers acquired electrical service in the 1920s and '30s, but most rural dwellers were not yet served by utility lines. *Popular Mechanics* addressed the electrical needs of rural families on many occasions over the years. A 1938 article called "Electric Plants Offer Power to Everyone," noted the availability of wind-powered chargers that could generate enough juice to run a radio and two lights. Meanwhile President Franklin Roosevelt's New Deal administration was making a concerted effort to electrify rural America by establishing the Tennessee Valley Authority to provide hydroelectric power to a large portion of the South, and by establishing the Rural Electrification Administration in 1935.

JULY

25 CENTS

POPULAR MECHANICS
MAGAZINE
WRITTEN SO YOU CAN UNDERSTAND IT

REG'D. TRADE MARK, GREAT BRITAIN. No. 410426

REG. U.S. PAT. OFF.

PAGE 8

Brightly lit cityscapes were rapidly becoming a point of civic pride in 1932, when "Building with Light" was the inspirational theme of this *PM* cover.

Industrious readers could invest muscle, rather than money, in entrepreneurial projects. This 1935 feature noted that "electric lights are an added inducement to tourists."

"Blueprint for Tomorrow"

Throughout both the Great Depression of the 1930s and the wartime shortages of the 1940s, *Popular Mechanics* presented a telling combination of fantastic and functional articles. The magazine optimistically offered readers hope for a better future through technology, while simultaneously offering a course of practical action—things readers could do to improve their own circumstances and those of their country.

As one might expect, money-making projects abounded. "Tourists' Cabins that get the business" were promoted in 1935 as an ideal project for readers located along popular travel routes, affording "substantial income in the summer months." Entrepreneurial efforts were boosted with "Effective Signboards for the Roadside Merchant," featuring ideas for signs selling everything from eggs to antiques to riding lessons. Articles on sports and inexpensive craftwork dominated other, more costly hobbies. The reuse of cast-offs and the repair of broken items around the home, shop, and farm had long been a part of *PM*'s mechanical lore, but the 1930s brought a renewed emphasis on how-to articles like "Repairing Electric Appliances" (1934), as well as a regular illustrated feature called "Solving Home Problems."

Alongside its continuing emphasis on the practical present, the magazine nourished the dream of a fanciful future. Colorful covers depicted shining cities pierced by smoothly flowing ribbons of ground and air vehicles; massive airships of the future full of well-dressed travelers; spacious, curvilinear office towers; and speeding bullet-shaped trains, boats, planes, and cars. Inside the magazine, designers and engineers painted a glowing future. Dozens of articles proclaimed that new industrial processes for using such materials as magnesium, aluminum, rubber, and porcelain enamel held the key to industrial recovery in the world of the future. With these materials, industrial designers would mint a world that was the ultimate

in comfort, efficiency, and style by shaping every aspect to meet a practical or aesthetic need. In a lengthy article called "Planning the World of Tomorrow" (1940), the famous industrial designer Walter Dorwin Teague stated:

Our better world may be expected to make equally available for everybody such rare things as interesting, stimulating work, emancipation from drudgery and a gracious setting for daily life, freedom of movement, free exchange of thought, bodily well-being and mental equanimity.

Yet under this shiny veneer, *Popular Mechanics* articles revealed a high level of anxiety about the changing job market. "Has Radio Reached Its Peak?" asked one headline in the February 1934 issue. "Machines—Masters or Slaves?" asked another article debating the costs and benefits of mechanization. But the answers that followed were reassuring— even inspirational, "Science is at our service—it can enslave us, or it can free us—but we ourselves must make the choice." Many articles encouraged readers to get training in fields that promised continued or new growth. One 1934 article titled "Does Opportunity Still Exist?" named radio, television, movie production, aviation, mining, auto design, and research as industries offering "New Opportunities Which Did Not Exist a Few Years Ago."

Perhaps most characteristic of the attitudes expressed in *Popular Mechanics* issues of the 1930s was the strapping male giant striding across the magazine's 30th anniversary cover, clothed only in a banner announcing "PROGRESS." At his feet, looking optimistically

Tinkerer, mechanic, designer, inventor, writer, and promoter, Bill Stout was the ultimate "popular mechanic." Well known for his design of the famous all-metal Ford Tri-Motor aeroplane (better known as the "tin goose"), Stout drew on his extensive background in aircraft design to conceive a revolutionary automobile which he dubbed the "Scarab" because of its aerodynamic, beetle-like shape. Several versions were hand-built between 1932 and 1936. Though never mass-produced, the Scarab received extensive press coverage. Stout himself wrote an article for *PM* promoting his "car of the future." Unusually roomy, with a space frame, no chassis, a rear-mounted Ford V-8 engine, moveable seats and a folding table, the car anticipated the look and function of today's mini-vans.

upward and forward, stood the hope of the modern age—a scientist in a lab coat, a laborer with his tools, and a suited businessman. Dozens of men representing those three fields wrote short pieces for that issue with titles like "Wonders To Come," "New Industries To Rise," "Luxuries For Everyone," and "The Past Assures The Future." Clearly Americans were no longer as awestruck as they once had been by the revelation of new inventions. They now took

Out of the Air --- TOMORROW'S CAR

such development for granted. In 1932, Winston Churchill wrote an article for *PM* entitled "Fifty Years Hence" in which he noted:

We all take the modern conveniences and facilities as they are offered to us, without being grateful or consciously happier. But we simply could not live if they were taken away. We assume that progress will be constant.

Indeed, by the 1930s the word "progress" had become almost synonymous with technical

Predicting the future became something of a mania in the 1930s, as people seized on any forecasts of brighter times to come. This 1935 cover, projecting an efficient and beautiful New York City for 1960, accompanied an article titled "Do Prophecies About Inventions Come True?" *Facing page:* First developed to cut the wind-resistance of aeroplanes, streamlined design become the emblem of modernity in the 1930s, surfacing in products ranging from trains to teapots. "Planning the World of Tomorrow," by industrial designer Walter Dorwin Teague, prompted this 1940 cover depicting redesigned rail travel.

progress. Although social and cultural progress were expected to follow in the wake of improved technology, they were not the primary goal of inventive effort.

Intimations of war began to surface in the pages of *Popular Mechanics* in the mid-1930s, just as the arms buildup prior to World War I had been apparent in the early teens. Great changes—both technological and social—seemed to be in the offing. A 1941 article, for example, noted that the "picturesque, lone inventor" had slowly been replaced over the last few decades by large research laboratories in which many people contributed "anonymous parts" to the development of new materials and inventions, a method of inventing that offered distinct advantages in wartime. And on the eve of America's entry into World War II in December 1941, physicist R. M. Langer's article on the uses of uranium, "The Miracle of U-235," predicted that with the coming development of nuclear energy, "We can look forward to universal comfort, practically free transportation, and unlimited supplies of materials." In Langer's future world, do-it-yourselfers would not be excluded from atomic glory:

There is no end to the practical applications that amateurs can work out once the energy source is available. For instance, it may become practical to melt our highways instead of building them as we do now.

Almost immediately, however, editors across the country were asked to omit all reference to U-235 as the government instituted wartime censorship for security reasons.

HUNTING THE SECRET OF LIFE

POPULAR MECHANICS
MAGAZINE
WRITTEN SO YOU CAN UNDERSTAND IT

REG. U.S. PAT. OFF.

DEC.
25 CENTS
30c IN CANADA

SEE PAGE 808

"We are at the gateway to a new world."

"War makes incredible changes in our daily habits and modes of living, but history shows civilization always takes a step forward when peace comes . . ." claimed a reassuring article published in 1942. Bold, patriotic covers featured the Allies' best machines for defense and offense: snares to entangle enemy submarines or aeroplanes, and tanks, torpedoes, battleships, and planes of all types. Some covers urged readers to buy war bonds, others featured civilians—both men and women—at work for the war effort. But the emphasis on technology and its future remained central. A 1941 article captured the perfect union of the era's main interests: "Streamline Designs for New War Machines." Proposed by industrial designer George Walker, these machines were reportedly fashioned to increase the "striking power and mobility of America's fighting forces." But as the war began to draw to a close in 1945, more and more articles speculated about the consumer goods and idealized "Victory homes" that were bound to come.

How-to articles published during the war continued the Depression themes of fixing up, making do, and improvising with castoffs—but for different reasons. "Wood Bats for Metal Around Your Home Plate" suggested substitutes for materials needed for the war, and appliance repair became even more important as the production of many civilian goods all but ceased. *Popular Mechanics* additionally supported the war effort by offering plans for Victory gardens and civil defense shelters.

The rapid application of new military technologies to peacetime uses after World War II did little to disprove the theory that "civilization takes a step forward when peace comes"—at least in the technological arena. In 1945 readers avidly consumed information about the brand-new marvels of radar, which promised great improvements in air traffic control. Microwave technology, discussed in *Popular Mechanics* as early as 1931, was now being applied in many areas of industry, communications, and medicine. In 1946, the magazine noted that microwaves were also being used to cook food in ovens called "radio stoves"; ten years later, readers saw a photo of one of the first "microwave ovens" available for consumers. "It Thinks With Electrons" introduced the army's monstrous Electronic Numerical Integrator And Computer, ENIAC. The government had planned to use its 18,000 vacuum tubes to solve wartime ballistics problems, but now computer power would be applied to peacetime needs as well.

While most new wartime tech-

Physicist R. M. Langer tirelessly promoted atomic energy in the early stages of its development. The illustration accompanying his 1941 *PM* article captured Langer's optimistic dream of a world liberated by clean, as well as limitless, power.

Above, author's ideas of what life may be like in uranium age. Note that most activities are located underground—even farming. Only transportation and recreation are above the surface. Left, Dr. Langer's conception of the U-235 automobile

Left: This 1944 cover revealed the innovative purpose of the odd-looking prow of the American LST (Landing Ship Tank) widely used during the war.
Right: George Walker's 1941 proposal for a streamlined torpedo boat.

nologies were welcomed into peacetime use by the postwar public, the pros and cons of atomic applications dominated the big picture of science and technology presented in the magazine for many years. The possibility of practical atomic energy had been discussed in *Popular Mechanics* as early as 1923, and again in the early 1940s, with wild optimism expressed about its potential. But nothing had prepared readers for its use in a bomb. One writer in the November 1945 issue described the world's reaction to the A-bomb this way:

There were three atom bombs and four explosions: New Mexico, Hiroshima, Nagasaki, and, the fourth and most shattering of all, the blast that plummeted the

minds of untold millions into the strange new world of atomic energy.

Popular Mechanics had often described the dangers and problems of developing and using technologies, but at no time were the tradeoffs demanded by technology more clearly recognized than in the postwar years. A 1945 article called "Bringing the Atom Down to Earth" captured the polarities of the debate:

The fact is that the dreaded but alluring atomic age is already here. Whole nations have grasped that fact almost with the speed of light. They may cringe before it—but they want it for its advantages and in spite of its dangers.

The magazine soon brought the nuclear age home, publishing

plans for do-it-yourself bomb shelters, and speculating about the underground cities that might be necessary after an atomic war. Readers could cash in on the " '49 uranium rush"—spurred by the Atomic Energy Commission's reward for uranium-rich ore samples—by building a portable Geiger counter from *PM* plans. In 1955 the magazine followed up with a story about a California reader who hit pay dirt using his *PM* Geiger counter, and ended up owning major interest in a promising uranium mine.

Articles on peaceful uses of atomic energy, however, also raised the specter of danger. Prewar dreams of atomic power had been optimistic but by 1946 a more realistic view emerged:

The chances are that you never will drive an atomic automobile or fly in an airplane powered by nuclear energy . . . . The trouble is that atomic power packages radiate vast quantities of deadly, invisible rays.

Though largely optimistic about developing practical and safe applications of atomic energy, most postwar articles cautioned that development would be slow and future applications difficult to predict. By the mid-1950s, atomic developments and discoveries were occurring fast enough to fill a regular column titled "Keeping up with the Atom." Published from 1955 to 1961, it noted both the positive and negative aspects of atomic applications. In 1960, in one and the same column, a *PM* editor reported: "Daily, in our reactors we are creating monsters that must be buried and then watched for centuries," while "On the credit side, the atom is being used to produce a highly improved polyethylene film." (*See Michael L. Smith's essay, page 110.*)

"Popular Shopping"

In dramatic ways, the end of World War II signaled the completion of *PM*'s move to becoming a modern consumer magazine. Even as its readers explored the global implications of new technologies, they became increasingly involved in domestic comforts, recreation, and the consumption of the ever-growing variety of machines and gadgets on the market. In 1941 the magazine published its first "Where-to-Buy-It" index for consumer items. Soon after, *PM* dropped its long-

Below left: With materials in short supply during the war, *PM* often reported on creative substitutions devised by manufacturers and readers. *Below right:* Despite *PM*'s fondness for predicting the future, this 1946 coverage of the first electronic computer gave no hint that computers might ever do anything but perform "extremely complicated calculations."

established policy against including manufacturers' names in its stories, and began to publish product comparisons for cars, tools, televisions, lawn care products, hi-fi equipment, and appliances. Stories about farm machinery and trade skills, once found in almost every issue, had declined steadily after World War I, and had almost entirely disappeared by the 1950s.

Just as after World War I, returning veterans spurred a postwar housing boom, a baby boom, and a consumer boom. With the purchase of houses made possible for veterans by low-interest housing loans from the government, model "dream homes" were of great interest to *Popular Mechanics* readers. The magazine responded by offering dozens of detailed, build-it-yourself house plans featuring all the most up-to-date conveniences. The magazine also responded to America's fascination with cars by initiating the annual auto issue. In fact, *Popular Mechanics* could claim to have codified the essence of postwar American consumption in the annual auto, boating, outdoor living, and housing issues it has been publishing since the 1950s.

The 1950s were a boom time for American families, the economy, and for *Popular Mechanics*. Letters to the editor were first published regularly beginning in 1949, and soon proud readers were using the new forum to show off the things they built from *PM* plans. Clearly *Popular Mechanics* readers welcomed information on consumer goods, especially those suitable for fixing up their homes. This may, in fact, have accounted for the magazine's dramatic increase in circulation—from about 750,000 in

In the post–World War II period, the country faced an acute housing shortage. After the deprivations of the Depression years and the war, many Americans wanted nothing more than to own a home. Single-family housing starts skyrocketed from 114,000 in 1944 to 1,692,000 in 1950. *Popular Mechanics* devoted many postwar issues to "dream houses," particularly ones its handy readers could build themselves. The up-to-date ranch home on this 1951 cover was built by novices Tom and Vinita Riley of Portland, Oregon, from *PM* plans in Tom's words, "to prove that if I could do it, anyone could."

1945 to over a million in 1946. Advertisements—now interspersed among the editorial pages—featured tools and home improvement items, cars and appliances, cigarettes and spirits: clearly *Popular Mechanics* was a producer of consumers as well as a consumer product.

Those consuming the magazine, however, were no longer just Americans. During World War II, American servicemen abroad had exposed Allied soldiers and others to copies of the magazine. Immediately after the end of the war, a French war hero, named Adrien Albarranc, made his way to Chicago to inquire about publishing a French version of *PM*. It would be the first of several foreign language editions; by 1952, there were also Spanish, Swedish, and Danish versions. The magazine had always covered international developments in technology and science, now it would reach an international audience as well.

New Hands at the Helm

The only change in ownership in the long history of *Popular Mechanics* occurred in 1959. That year the Hearst Corporation purchased the magazine from H. H. Windsor, Jr., the son of *PM's* founder, and Editor-in-Chief since his father's death. Soon its headquarters were moved from Chicago to New York City. While there was no drastic change in the content of the magazine, there was a noticeable change in the look of the covers. Since 1911, *Popular Mechanics* had featured full-cover, color paintings of various technologies and activities, and in the 1930s and '40s these had become

The clutter of this Vietnam War cover did not entirely obscure the message of its artwork. U.S. government studies in night vision devices began in 1952, intensified in 1961, and in 1965 resulted in the development of this scope, which used fiber-optic lenses.

increasingly stylized. The 1950s covers maintained the colorful artwork, but featured domestic and recreational scenes in greater numbers. The trend toward domesticity continued into the 1960s, but the characteristically bold covers nearly disappeared as words increasingly competed with images: One cover had 136 words on it—not counting *"Popular Mechanics."* Photography again appeared on covers for the first time since 1909,

adding a touch of realism to many issues.

Cover art accurately reflected the content of the magazine. Throughout the 1960s and '70s, most articles focused on what was possible today rather than what had potential for tomorrow. Even "Man's Greatest Adventure" to date—the landing of Apollo 11 on the moon—failed to knock a photo of a pickup truck camper off the July 1969 *PM* cover.

Another sure sign of practicality was the increasing emphasis on consumer rights and safety. Ralph Nader was becoming a household name as *PM* articles addressed such topics as quack medical devices, deceptive product packaging, seat belts, air bags, and speed limits through the 1960s and '70s. Many of these articles revealed some skepticism about various technological developments. A 1964 article, promising to explain "Here's Why Those Do-Everything Lasers Don't—Yet," began with the comment:

With typical American overstatement, the laser has been touted as another instant miracle. (Remember the wild predictions about the coming wonders of atomic energy?)

A 1963 article used humor to address technological anxiety. Illustrated with cartoons of computers displaying human characteristics, it pointed out with a combination of relief and glee that computers—like their programmers—were indeed fallible: "If you envision a world free of human error, a society that is regulated by the calm clicking of an infallible computer, you are in for a disappointment."

Despite occasional apprehension, computers were becoming the biggest news in home electronics since television. First transistors, then integrated circuits had increased computer capabilities exponentially while decreasing their cost. A husband and wife who had designed, built, and programmed a computer to take on household tasks, such as bookkeeping, were the topic of a 1968 article, "A Computer in the *Basement?*" By 1976 articles like "Home

LOOK WHAT THOSE KNUCKLE-HEADS ARE DOING!
(MACHINES, THAT IS)

2+2=7 RIGHT! WRONG!·2·2·9 PERFECT

ENGINEERS tout their mechanical brains as being superior to humans, but even the best of the robots make mistakes

From the early days of the magazine, occasional articles addressed readers' fears of being displaced—or in this case replaced—by efficient machines. Early in the computer era, this *PM* article attempted to reassure nervous readers that computers were not as all-powerful as they might imagine.

Sweet Computerized Home" and reviews of computer kits were appearing with regularity. Soon men who would never have dreamed of touching a typewriter were pecking away at electronic keyboards. Despite growing enthusiasm, however, predictions for the new fad were cautious:

Whose imagination doesn't run at least a little wild at the thought of buying a computer for a few hundred dollars that can do more than a million-dollar computer system of just a few years ago?

It's a fantastic dream. But what is the reality, in late 1978?

The 1960s saw the beginnings of a growing national concern about environmental damage. Soon *PM* articles were discussing air pollution, pesticides, nuclear waste, and dwindling natural re-sources. "Environment" became an index heading, and "Pollution Fighters' Newsletter" a regular column. The public now had a catch-all term—"alternative"—and a somewhat heightened consciousness that brought experimenters in from the fringe. Wind and sunlight were certainly not new sources of power, nor new to *PM*, but they had now acquired a renewed cachet. The oil crisis of the 1970s gave added impetus to one of *PM's* all-time favorite topics: the small, efficient personal vehicle, now labeled the "commuter car." A 1974 article in the magazine titled "A Gasless Way to Go" asked "Can you really beat the fuel shortage? If you're willing to exchange leg power for long waits at the gas pump, you may be able to—at least to some extent."

AUGUST 1984 $1.50

Popular Mechanics

1½ MILLION-MILE REPORT
Dodge/Plymouth Minivan Owners: "Quality's Great!"

Jimmy Carter: How To Handcraft Country Chairs The Old-Fashioned Way

COMING: Cheap Cars From Korea

PLANS: 5 Home Phone Centers For All That New Gear

CAR STEREO: Best Systems In 3 Price Ranges

STEP-BY-STEP: 5 Charming Garden Pools You Can Make

MAN-MADE ISLANDS: How We'll Build Them And Live On Them

Ex-President Carter shapes chairs from hickory grown on his own farm.

Once the most common method of producing goods, hand craftsmanship is today more often a hobby for Americans. Longtime craftsman and former President Jimmy Carter made the cover of *PM* in 1984 when he used traditional hand tools to fashion several chairs from green hickory.

"Technology is already overtaking our dreams."

If Americans were less fascinated with the future in the 1960s and '70s than they had been in the past, it was perhaps because the technological development of the present was so rapid and so startling in its impact. Articles like "The Dream Comes True," on the Apollo 11 moon landing, reflected the feeling of many people that once impossible dreams were becoming reality with mind-boggling speed.

Continuing development in electronics remained a major theme into the 1980s, as microchips expanded the capabilities of every manner of machine from space vehicles to microwave ovens. The laser, once called "a solution in search of a problem," was now creatively employed in a variety of tasks: as a tool for eye surgery; as the "stylus" of the new compact disc player; and as a cash register price scanner. Consumer articles occasionally expressed a certain weariness from the effort to keep up with new developments, but somehow managed to maintain their usual enthusiasm. A 1971 article on cartridge television systems (forerunners of today's videocassette systems) ended by stating: "If the array of choices seems to offer hopeless confusion, there's one compensation—it's the most exciting confusion to come along in ages."

As we all know, the use of microchips made the "new stuff" more difficult to explain and understand, let alone fix. Many of the new devices came to be known as "black box" technologies, referring to the sealed cases that housed delicate miniaturized circuitry. Many electronic consumer products carried labels recommending they be returned to the manufacturer for service. In 1982, one *PM* editor asked his readers "Do your eyes glaze over at the mention of microcomputers? If so, you're not alone." Car owners and garage mechanics complained to *Popular Mechanics* automotive editors about the mystifying intricacies of new cars. Undaunted, *Popular Mechanics* kept up its tradition of providing consumer advice and how-to information. A 1989 car care article, "12 Tools for Fixing Your Eighties Car," attempted to reassure resentful do-it-yourselfers:

Restoration buffs give many reasons for their hobby—from nostalgic longing, to investment, to preservation of craft skills or technological history. It may be more than a coincidence, however, that increasing numbers of enthusiasts—like the subjects of this article—are embracing older, more familiar technologies as electronics revolutionizes present-day technology and modes of work. *Above:* Years after modifying a 1950 Schwinn Black Phantom bicycle in a burst of youthful enthusiasm, its owner had it fully restored.

Two of America's most characteristic enthusiasms—automobiles and the future—are embodied in "concept cars" like this 1986 Ford Probe V. Built since the 1930s as a way to develop new ideas and enhance company image, concept cars, like most experimental technologies, capture their builders' vision of the technological future.

No question, it was tough sledding for even the best Saturday mechanics [in the 1970s.] But the cars of the late '80s—not to mention the 90s—are definitely improving from the DIY point of view.

Soon many hands-on enthusiasts began to turn for comfort to older forms of technology. Although old machinery had always sustained a small group of fans, the mid-1970s saw the beginning of a growing trend toward restoring, collecting, and using old cars, houses, engines, trains, aeroplanes, and furniture. Articles on traditional craftsmanship began to appear, and the restoration of national monuments such as the

Statue of Liberty was covered in feature articles. Starting in 1975, the covers of the annual car care issue displayed beautifully maintained classics of decades past. In 1986, "Old House Restoration" became a regular column. *PM* even looked to its own past beginning in 1984 with a regular feature titled "*PM's* Time Machine."

Reasons for America's "restoration fever" occasionally surface in passing comments published in the magazine. In 1989, "Recycling the Classics" cited "nostalgic longing" and "lost youth" as reasons enthusiasts restored old bicycles. The satisfaction of working with tools and machines of familiar (but by no means simple) con-

struction has become a strong motivation as a counterpoint to coping with the electronic systems of today's world of work. Former President Jimmy Carter, in an article on his traditionally crafted hickory chairs, expressed the feelings of many of America's hands-on history buffs when he said: "I really find manual labor a great release—a kind of vacation for me—something that I hungered for when I was in the White House."

While interest in older crafts and machines continues today, the pages of *Popular Mechanics* are beginning to show a resurgence in curiosity about the future. Concept cars, those futuristic dream machines so characteristic of the 1950s and '60s, have staged a return in the auto industry, and correspondingly have reappeared in *PM*'s pages. Speculation about future technology is more popular in the magazine than it has been since the 1930s and early '40s. Covers of the last few years are again featuring topics like "Diver of the Future," "Tomorrow's Trucks," and "Futuretrain." The superlatives so common in *PM*'s earliest headlines have also reappeared in such headlines as "The World's Biggest Planes" and "Navy Builds The World's Largest Blimp." Dramatic paintings again predominate on covers, and their stylized and spare look is evocative of the streamlined covers of the 1930s. While enthusiasm for technology may have had its ups and downs over the last 90 years, it is clearly not dead—indeed the pendulum of enthusiasm may once again be traveling forward.

As the spectrum of technological applications has grown throughout the 20th century, so has the possibility for different expressions of enthusiasm at varying levels of interest. The stereotypical enthusiast—and perhaps the stereotypical *PM* reader—is someone with an abiding love of machinery, a desire to know how things work, and the ability to build and repair things. That dedicated tinkerer may have dominated the universe of technology at the turn of the last century, but today there are many other types of technological enthusiasts. Consumers who have no idea how modern appliances actually work can nonetheless enthusiastically compare features, purchase, and plug in. Thousands of children find amusement in video arcades and on home computers. And millions of people owe their livelihood to technological production, or use modern machines daily to accomplish their tasks. Even people who claim to dislike "technology" are often dependent on it and enjoy its benefits. For instance, most people abhor the pollution caused by coal- and nuclear-fired electric power plants, yet are utterly dependent upon—and appreciate—the end product. To one degree or another they are all enthusiasts.

Popular Mechanics caught a wave in 1902—a growing wave of enthusiasm for the machines of our times. Technologies being developed around the world were changing people's lives in unpredictable ways, making life simultaneously easier and more difficult; safer, yet more dangerous; more independent in some ways and more dependent in others. In the course of its 90-year history, *Popular Mechanics* has recorded abundant evidence of diverse dreams embodied in equally diverse machines. The popularity of technological literature across this century attests to a widespread desire to understand technological change and turn it to personal advantage—evidence of our dreams of building a good life and a better world by using technology.

Public concern about environmental issues, heightened in the 1960s and '70s, remains apparent. In 1990, "Stripmining the Sea" examined fish poaching, driftnetting, and international regulations.

JUNE 1991 $1.95

Popular Mechanics

WORLD'S FASTEST PLANE

NASA Calls The SR-71 Blackbird To Active Duty

WHAT WE LEARNED FROM DESERT STORM
America's High-Tech Killer Punch Has Changed The Way We Win Wars

SPORT-TESTING BIG-BUCK SNEAKERS
We Compare The 5 Hottest Models For Value, Performance And Class

RESTORATION DREAM
Build A Traditional Covered Porch

© Anita Hopper

08638

0 754744 1

06

Flying with an Easily Constructed Glider

Belmont, California.

Editor Popular Mechanics Magazine:

I thought that you would be interested to know that a glider, patterned after the plans which you published in your magazine, has been made and was very successful. The accompanying photograph is a picture of me as I was leaving the top of a hill. I could not get more than thirty feet above the ground, nor travel more than a hundred yards, because the hill was not steep enough. All of the better hills around here were so covered with brush that it was impossible to try flying from them.

DE RONDE TOMPKINS.

Technology in the Household: A Reminiscence

JOHN L. WRIGHT

My father, born in 1913, grew up on a central Illinois farm without electricity, telephone, or indoor plumbing. As a boy, he learned to plow and cultivate the fields with horse-drawn implements. My son, born in 1976, is growing up in a city in a world of high technology. He spends much of his time watching cable television, playing complex games on the computer, listening to compact discs, talking on the telephone—sometimes doing all of these simultaneously.

The generations—my father's, my own, and my son's—span almost all of the 20th century. Our differing assumptions about everyday life reflect the staggering changes wrought by technology in this era. Each of us has been shaped to a great degree by the technological environment of his time. Technology in the broadest sense has been much more than simply a set of tools to perform the necessary tasks of living. Technology has defined our universe of thought and action. It has structured our lives in every way: how and where we live, how we get educated, how we earn a living, how we communicate, travel, buy and sell things, and how we spend our leisure time.

Every individual's perception of

The author's modernistically styled boyhood home in Jacksonville, Illinois, as photographed shortly after its construction in 1935.

technological needs at a particular time and place is largely determined by the technology reasonably available, i.e., "you don't know what you're missing if you've never had it." As my father studied his schoolbooks by the light of a kerosene lamp, he and his farm family did not feel deprived, even though many cities had already had electricity for close to 50 years. My son, on the other hand, is desperate when the hard disk on the computer crashes and is out of order for a day or two. That we so quickly come to take new technologies for granted indicates the pervasiveness and transparency of

the impact they have on us.

My father was a faithful reader of *Popular Mechanics* for some 40 years, from the late 1940s to the late 1980s. It was his primary source of information and advice on all sorts of matters, from what tools to buy to how to put up wall paneling, fix a leaky faucet, or sharpen a saw. It was one of the few magazines coming to our house that was exclusively for him. A practical aid, it was also one of his links to the rural world of handcraft and self-sufficiency he had left behind. The magazine helped him cope with an increasingly complex world of technology.

The Television Age

When I was growing up in the 1940s and 1950s, my family experienced the advent of commercial television. In 1951 we got our first TV set, a big blond console model with a 21-inch black-and-white screen. Living in western-central Illinois, far from any big cities, my family and several of our neighbors chipped in to buy a tall antenna to receive flickering signals from network stations in St. Louis, 100 miles to the south. Where radio had long been a friendly guest in our home, television invaded with an electronic presence impossible to ignore. We were instantly connected with the world at large through pictures and sounds, however grainy and static-ridden they might be.

For many of us, television was the ultimate technology. Like many technologies with great power, it was at once mysterious and commonplace. Its technical workings, incomprehensible to the average person, made it all the more magical. Yet its content, its message, from professional wrestling to situation comedies and the world news, was mundane, if not banal.

To a child of the 1950s, it seemed that we lived in a Norman Rockwell world of small-town ironies, where the greatest sorrow was a skinned knee and the greatest joy was a trip to the ice cream shop. Of course, the adults in that world were dealing, as adults always do, with realities of money, politics, illness, and death. But for most children and adults it was the Golden Age of American technology. The boom years of the post-World War II era brought a seemingly endless stream of new

A mid-1950s Christmas greeting card proudly displayed the family's television set, flanked by the author and his sister Susan.

goods, a cornucopia of wonderful things clamoring to be bought on credit and hawked every night on the television set.

The Home

My family was middle in every way: middle-class, middle-income, middle-sized, living in a middle-sized town in the middle of the Middle West. Our aspirations were to fit in, to be comfortable, to be regarded as hardworking, decent, and respectable. To be part of this newly emerging postwar middle class required possession of a cluster of material goods: a nicely furnished suburban-style house with a well-kept lawn, a recent model car, and a variety of household appliances. The good life was defined in large part by a level of ownership of material goods that would help make life comfortable, safe, and predictable. After the difficulties and deprivations of the Depression era and World War II, most Americans were starved for new goods, and many now had the money or credit to buy them.

Technology was creating an abundance of new goods, and most Americans were happy to be consumers in this apotheosis of the consumer culture. We were technological enthusiasts through and through, although we hardly thought of it that way at the time. Of course from time to time we worried about the A-bomb, but we were assured that the nearest designated "fallout shelter" would keep us safe from harm. Comfort, convenience, and a modest sense of luxury were the watchwords of the day. Acid rain, global warming, or the need to recycle were concepts of the future. For us it was a buy it, use it, throw it away society.

Having a home of one's own was perhaps the key ingredient in the formula of success. The very idea of home ownership, as opposed to simply having a house to live in, implied a different kind of involvement or commitment to the demands of domestic life. Beginning in the later part of the 19th century, the ideal of the single-family small-town or suburban home had been at the center of American middle-class aspirations. After all, the home represented stability, success, and self-reliance. And for those who looked back to their roots on the farm, it perhaps symbolized the individualistic spirit of the pioneers who built log cabins on the frontier.

Within our 1950s home, some spaces were defined primarily according to categories of usage: the bedrooms for sleeping, the hallway for greeting visitors, the combination living room/dining room for family gathering and television watching, and the den for reading or kids' play. Our house, built in the 1930s, was already somewhat out of date by the 1950s, as it lacked the large recreation room or family room popular in the new ranch houses of the era.

Many of the spaces were defined not just by function but by gender. As with generations before, the kitchen was ruled by mother's hand, and anyone who set foot there was in her territory. However, there were also spaces reserved for the men of the house, the workshop or workbench, usually found in the basement or garage.

The basement as an actively used space was very much part of the plan for the up-to-date middle-class home. In earlier times the basement was known as the cellar—an often forbidding, unfinished, and dingy place used to store coal and canned food and to keep the clotheswasher-wringer. In the 1950s it was often transformed into the newly popular recreation room or home workshop.

The Home Workshop

When she was asked my father's whereabouts, my mother's response, "he's down in his workshop," carried a host of meanings to me and my sister. The workshop was his place and his only, just as the kitchen belonged to my mother. My mother, my sister, and I were visitors to his domain. This was understood as a masculine space. It was all lumber and hardware: tools, nails, paint, and house parts such as springs and door latches. It smelled of turpentine and freshly sawn wood. It was the place where things got fixed.

To be in the basement was to be down in the raw guts of the house, where you could look up to see the first-floor joists and listen to the low roar of the furnace. The basement and the workshop were places of facts and realities. No room here for the ambiguities of philosophy, literature, or polite conversation. Measure a board down to an eighth of an inch, cut it right, and it will fit. Do it wrong,

and it won't fit. No need to talk much about anything. The task is there—just do it.

Most of the other space in the house belonged to my mother. The kitchen, the bedrooms, the hallways, the living room, and the den were all finished, tidy, respectable spaces. Only the basement and the garage belonged to my father. Here a man could get his hands dirty and not worry about making a mess. The workshop was a masculine refuge in an increasingly feminized household.

Our home's division of labor by gender was followed strictly and without question. Mother did the shopping for food, clothing, and household odds and ends. She was responsible for cooking, cleaning, child care, and clothes-washing. Father was responsible for anything technical or mechanical, anything calling for repair or improvement. This included the house itself, the yard, the garage, the miscellaneous household tools and machines, and, of course, the car. If he couldn't fix the problem himself, he was expected to arrange to get it fixed by a repairman or mechanic.

The home workshop of the 20th-century middle-class house has usually been thought of in the context of male hobbies: building a birdhouse, a knickknack shelf, or puttering with some other pastime. Unquestionably this was part of its function. Many men like my father, who had grown up in a rural setting, were the first generation in their families to hold white-collar office jobs in town. The home workshop provided an outlet for their desire to work with their hands as well as their heads. This was therapy through manual

Remodeling of the family home in the late 1950s converted the original garage into a breezeway and introduced a carport. Redwood siding and another garage were added later.

The big self-propelled lawn mower was an important household machine and a favorite prop for family pictures. *Left:* The author and his sister share the seat in the early 1950s. *Right:* The author's son and father do likewise almost 30 years later.

activity—an anodyne to the "Age of Anxiety," as W. H. Auden called the mid-20th century. Many of these men looked at their workshops as a recreational escape from the pressures of everyday life. These places and activities served for some as a link to the rural world of the past, a world of the craftsman and artisan.

At the same time, the home workshop served a quite practical function for the new middle-class man. Unlike many of their counterparts of the late 19th and early 20th centuries, most middle-class families of the mid-20th century did not have the luxury of servants or household help. Thanks in large part to government housing loans, the millions of new post–World War II homeowners had just enough money to meet mortgage payments. Few had enough left over to hire help to maintain and repair their houses. Thus was ushered in the great age of do-it-yourself-ism.

For my father, home ownership held a deep obligation to keep the house, the yard, the car, and all the other family possessions in good order. A great deal of his "leisure" time was spent on the never-ending tasks of maintenance and repair. As any homeowner knows, sheer survival requires an eternal battle against the forces of decay and disorder: windows get broken, doors jam, gutters fall off, crabgrass spreads, and appliances break down. There is no rest for the conscientious home owner.

Many home owners also launched do-it-yourself home remodeling projects. As families grew and more living space was desired, the bungalow that had

John B. Wright, the author's father, subscribed to *Popular Mechanics* for 40 years. Many of his projects for the home were carried out in his workshop.

seemed big enough a few years ago now needed additions and modifications. Changing house styles and newly available materials spurred home improvements.

The Popular Mechanic

It is within this context of recreation and necessity that technology became domesticated into everyday life. At one extreme, technology was embodied in massive systems such as electric power generation. At the other end of the scale, and equally important in its own way, was the homeowner replacing a blown fuse or repairing a lamp. Many technological systems depended, and continue to depend, on both professionals and amateurs for their maintenance and use.

My father and *Popular Mechanics* parted ways just a few years ago, as many of the household tasks became physically difficult for him and as the magazine became more oriented to the consumption of new technologies. But the magazine and the world of know-how it represents are part of our memories, an icon of a more naïve and exuberant era of American enjoyment of technology.

JULY 1953   35 CENTS

POPULAR MECHANICS
MAGAZINE
WRITTEN SO YOU CAN UNDERSTAND IT

Drag Racing
page 131

WHAT OWNERS SAY ABOUT
the CHEVROLET

For the Craftsman
MAKE MONEY CUTTING
SEMI-PRECIOUS STONES

Popular Mechanics, July 1953. The cover story was an article by *PM*'s Western
Editor, Thomas E. Stimson, Jr., titled "Lightning Hits the Drag Strip."

Strip, Salt, and Other Straightaway Dreams

IN THE QUEST FOR STRAIGHTAWAY
SPEEED, MOMENTUM HAS TYPICALLY
BEEN IMPARTED BY AN UNABASHED
ENTHUSIASM FOR "SETTING THE
RECORD." THE SEA CAPTAIN WHO
DROVE HIS CLIPPER SHIP
1,478 MILES IN FOUR DAYS
AND THE HOT RODDER WHO
DESIGNED A GO-CART CAPABLE
OF A 6.3-SECOND QUARTER MILE,
WOULD HAVE UNDERSTOOD
ONE ANOTHER PRETTY WELL.

ROBERT C. POST

Henry Ford once remarked that the outcome of a technological innovation is contingent upon three questions: "Is it needed? Is it practical? Is it commercial?" Perhaps this was true in Ford's world. But how are we to account for technologies that are pushed despite ambiguous replies to one, two, or sometimes all of these questions? Turn the pages of any issue of *Popular Mechanics* and you are bound to find machines to which momentum is imparted by a different question: Is it possible?

A 1931 study titled *Industrial Creativity: The Psychology of the Inventor* indicated that inventors cited "love of inventing" as their primary urge more often than they did either "financial gain" or "necessity or need." "Love of inventing" may readily be understood as "enthusiasm." "If we fail to note the importance of enthusiasm that is evoked by technology," writes the historian Eugene Ferguson, "we will have missed a central motivating influence in technological development."

One can find enthusiasts in any technological constellation, from 16th-century German clockmakers and their "materialized fantasies" to the computer whizzes who populate contemporary sagas like Tracy Kidder's *The Soul of a New Machine*. For their own reasons—which are different from the reasons put forth for political purchase—enthusiasts push things like superconducting supercolliders and manned space stations, Star Wars, and Stealth bombers. Had these same people been around two centuries ago, they might have been the ones who pushed mechanization: "When it became possible to make thread and cloth by machine," says Brooke Hindle, another historian who has sought to alert us to the import of technological enthusiasm, "they were so made; when boats, trains, or mills could be driven by steam, they were so driven."

It might be added that, in our own time, when it appeared possible the earth could be circled nonstop, an aeroplane called *Voyager* was invented, or when it appeared that a man might fly through the sky with only the power of his own muscles, the likes of the pedal-powered Gossamer Condor were invented. While the burden of funding is an entirely different matter, as far as motivation is concerned there is very little difference between creating a Gossamer Condor and a Stealth bomber. Neither seems to have had much to do with need or practicality and certainly nothing to do with commercial viability.

Vehicles have always been especially seductive to technological enthusiasts, and speed has been a frequent, if not inevitable, ideal. To be sure, other motives may be mixed in with sheer enthusiasm. The transatlantic packet, the Western steamboat, the high-wheeled express locomotive, and all such storied contrivances from the American past were aspects of a "technology of haste" that was linked, in Daniel Boorstin's words, to "rewards that others might grab if you were not there before them." The question "How fast will it go?" was one that gained currency as Americans became preoccupied

with ways of "saving" time (or in the case of clipper ships, with "clipping" it). "Getting there first" was closely connected to commercial advantage.

Yet, speed was sometimes idealized at the *expense* of commercial advantage. Prime but by no means unique instances of this were the windjammers and oceanic sidewheelers of the 1840s and early '50s; limited capacity and large operating costs made them a profitable proposition only within such a narrow economic context that freewheeling entrepreneurs were left beached when the context changed even slightly, as it would do a few years before the Civil War.

A century later, firms like Cunard and United States Lines were still chasing the "Blue Riband" for the fastest transatlantic crossing, even as the impending fate of pacesetters like the SS *United States* was foretold in the emergence of the Dash 80, the first of the 707 family of jetliners, and enthusiasts at places like Boeing were exploring the feasibility of flying passengers around at three times the speed of sound, Mach 3. And why not? Denizens of the Society of Experimental Test Pilots would reach speeds near *seven* times the speed of sound in the 1960s.

In other quarters, however, people were becoming more attuned to practical tradeoffs. Priorities were beginning to shift. As it turned out, no appeal to international gamesmanship was sufficient to make the Supersonic Transport (SST) materialize, and since 1974 even conventional jets have been throttled back. For Americans, speedy railroad trains

are pretty much a foreign phenomenon, and on the road there are those 55 miles-per-hour signs in places where the posted speed was once 65 or 70. Truckers habitually flout the law, of course, and the notion that time is money has also sustained sales of Learjets and the strategy of firms like Federal Express. But virtually no form of commercial transport maintains the schedules it could maintain if the costs were deemed worth the benefits.

Even so, enthusiasm for vehicles that simply "go" fast has never waned—particularly not in the U.S.A.—and with such vehicles Americans have usually stayed near the leading edge. Not that an enthusiasm for speed is a unique national characteristic, far from it. The Anglo-French Concorde first exceeded twice the speed of sound in 1970, a few months before Congress shot the SST, and despite extravagant costs Concorde flies transoceanic routes to this day. Italian names are synonymous with speedy cars, German names with the fastest bobsleds. Australians have always found speed exhilarating, and in the summer of 1983 they proved that the American monopoly on 12-meter sailboat racing was not forever by taking away the America's Cup. The so-called land speed record (LSR) was a British passion even before Sir Malcolm Campbell first bettered 300 miles per hour on the Bonneville Salt Flats of Utah in 1935.

Grand Prix, in some regards the most sophisticated of all forms of racing, has attained only a tenuous acceptance in the U.S. But that is precisely the point. Formula 1 cars are engineered for a broad

Two classic American hot rods, Ford roadsters stripped of their fenders, have it out in 1952 before a sparse crowd at the drag strip on the Los Angeles County Fairgrounds in Pomona, California.

range of functions, for braking and cornering, not just speed alone. Americans tend to favor their speed contests neat—in the case of autos, the lapping of super-elevated ovals that are virtually one continuous straightaway. Or, even more peculiarly, their machines actually *do* go in a straight line, ending up not where they started but in a different place altogether, just like a clipper ship or a prancing 4-4-2 locomotive.

With an Indianapolis 500 that draws one of the largest crowds of any sporting event in the world, Americans are clearly not indifferent to "closed course" competition, where the starting line is also the finish line. But along with the various ovals, loops, and road circuits there are hundreds of point-to-point courses, where the finish line, point B, is somewhere different from the start, point A. Sometimes B is a long way away, as in the competition that persists in storied affairs like the coast-to-coast Cannonball Run. Yet, for many enthusiasts, a test of endurance has nothing like the immediate appeal of flat-out speed. The course can be perfectly straight and drastically abbreviated.

Standing-start "sprints" had been staged in both England and Australia in the early years of the 20th century. But affection for straightaway racing grew strongest among a peculiar breed of enthusiast that flourished in southern California, young men dubbed hot rodders. Hot rodders first became famous, or rather infamous, in the late 1940s, as the newspapers discovered that they posed a public menace, and menacing images were embellished in the fiction of Henry Gregor Felsen and a peculiar genre of "B" movie. Menace or not, such enthusiasts had been around for a good many years, constructing "gow jobs" when Model-T Fords were new. Indeed, Henry Ford himself is as likely a candidate as any for designation as the first hot rodder.

Even before Bonneville became the prime venue for chasing speed records, the heirs of Ford's enthusiasm had been racing across the dry lakes of California's High Desert. The first organized event was held in 1931, at a place called Muroc. After Muroc was appropriated as part of a test facility for the Air Force and the National Advisory Committee for Aeronautics (it was later named Edwards Air Force Base in honor of a test pilot who died in a Northrup Flying Wing), the hot rodders were evicted. So, while Chuck Yeager was trying to break the sound barrier in *Glamorous Glennis* high above their old race course, they moved to El Mirage, a few miles from the town of Adelanto in San Bernardino County.

Events at El Mirage were held under the auspices of various loose-knit sanctioning bodies, foremost being the Southern California Timing Association (SCTA), which dated from 1937. By nature, dry lakes were dusty, and only the driver who was leading a race could see where he was going. The dire consequences were predictable. Hence the SCTA restructured

At Santa Maria, California, the start of the final round of racing between two 1953 dragsters; the bodywork is hand formed, but the running gear is mostly Ford.

the activity as "timing" rather than racing. In 1949, in addition to its events at "the lakes," SCTA began staging "Speed Week" each August at Bonneville, where competitors could get a longer run, several miles. The surface was hard as concrete and white as snow and dust was no problem. Even so, the activity remained "timing." As at El Mirage, cars were sent off one by one, and drivers headed straight towards a pair of photoelectric beams, "traps" that started and then stopped a timer. They were racing only the clock.

But "the clock" is, after all, an abstraction, and many enthusiasts got more satisfaction out of actually competing wheel to wheel with somebody else. Oval tracks were an option, but one that precluded the thrills of flat-out speed. Public streets were another option, but, needless to say, the dangers were multiplied. Although hot rodders may not really have signalled the new wave of juvenile delinquency, the newspapers not-

withstanding, plenty of them remember spending a night behind bars after getting nabbed while street-racing. Whatever other lessons they may have learned, however, they had also found out that they could test their mechanical ingenuity quite nicely in a short, straight sprint of, say, a quarter-mile, and get up a lot of speed as well.

So it was that drag racing was defined at the midpoint of the 20th century—the first commercial drag strip began operation in June 1950, at Orange County Airport in Santa Ana, California, and soon there were similar operations all over the state as well as in places such as Caddo Mills, Texas, Zephyrhills, Florida, and even Half Day, Illinois. One challenge was to attain as much speed as possible, but the short distance posed a serious limitation, so it became even more important to cover the ground as *quickly* as possible (as had been the case with clipper ships and express trains).

Now, hot rodding had fostered two different sorts of enthusiasts. The land speed racers preferred a solitary contest with an electro-mechanical rival, and the "numbers" they prized most were "m.p.h." They liked having at least a mile to get up to speed before tripping the first beam (as at El Mirage), preferably several miles (as at Bonneville). The drag racers preferred one-on-one racing at close quarters, with everything to be settled within a distance of 1,320 feet. Though not inattentive to "m.p.h.," they soon found that a machine capable of a fast speed at the 1,320 mark was not necessarily the quickest about getting there. Hence their favorite numbers were not a measure of *velocity* but rather of *time*, elapsed time from start to finish, "e.t."

At El Mirage and Bonneville, the fastest cars were geared so that they lugged slowly away from the starting line, usually needing an assist from another vehicle, and did not really get moving until almost out of sight; they could be heard long after they had disappeared in a cloud of dust or over the horizon. At the drags, they charged off furiously with tires smoking and engines screaming, and everything was over within a few ticks of the clock as parachutes blossomed to slow them down before they ran out of room. Drag racing seemed to be all a matter of brute power; on "the salt" or "the dirt" such touches as aerodynamic finesse were equally important, and, while speeds were higher, the pace seemed much more leisurely.

However different they were in their particulars, *Popular Mechanics* was naturally enthusiastic about both activities. "The hot rods have grown up," wrote Ewart Thomas in a 1950 article titled "Hot-Rod Derby on the Salt Flats." "Amateur automobile testing has become a recognized sport," he noted, "and most of the rods are pure racing machines of advanced mechanical design." In 1953, Thomas E. Stimson, Jr., who had also recounted how "Hot Rods Burn Up the Desert" the year before, described the way in which "professional mechanics and amateur mechanics use the drag strips to test . . . any improvement in performance that special racing equipment provides."

Technological enthusiasm was what imparted momentum in each case, and at the start there was a lot of overlap among the participants. Many of drag racing's early pacesetters had been lakes racers and most Bonneville standouts of the 1950s were also drag racers. But the two groups began to separate, one from the other, in the 1960s, and today they are quite distinct for the most part. Don Garlits, the most famous and successful of all drag racers, set a record at Bonneville in 1988, but that was just an exception to prove the rule.

In addition to El Mirage and Bonneville, there were only a few other places anywhere on earth that were suitable for the pursuit of the same activity as well as being readily accessible. Until recently the rank and file of hot rodders had access to the salt only one week a year. While there were usually a half-dozen lakes meets each year, they were run Sundays only and were usually wrapped up shortly after midday as the temperature began to climb beyond 100 degrees. There was no racing, only timing of individual runs. In contrast, by the 1960s there were

The Texas Motorplex, south of Dallas, a drag strip for the '80s. VIP suites flank the starting line, 18-wheel support vehicles are lined up in the pits at left. Compare the grandstands to those seen on page 101.

The first all-out run by Don Garlits's innovative *Swamp Rat XXX*, Gainesville, Florida, March 1986. Crewchief Herb Parks is behind the tire, enthusiastic daughter Donna at right.

hundreds of drag strips all across the country; weather permitting, they were in operation every weekend, and sometimes weekdays as well. Solo time-trials were merely a preliminary; racing two by two was the primary activity, with successive elimination rounds as in a tennis tournament.

There were two separate pairs of photoelectric beams, one pair at the start and finish for elapsed times, the other closely spaced at the finish for timing miles per hour. While the driver who set top speed naturally gained bragging rights, emerging as the "top eliminator" required winning several rounds of racing; both car and driver had to be quick. Drag races had something of the flavor of a Western shoot-out, whereas the denizen of the lakes or Bonneville more closely resembled a solitary marksman taking target practice.

There was an even more significant distinction between hot rodding's two primary offspring. The drags had a commercial tinge right from the start. Tom Stimson of *Popular Mechanics* called the founder at Santa Ana—C. J. ("Pappy") Hart—an "enthusiast," and he certainly was; but Hart was always thinking in terms of a moneymaking venture. As more and more spectators caught on to the excitement of "the fastest sprint cars in the world" (as another *PM* writer put it in 1959), certain racers, beginning with Garlits, began to regard themselves as pros. They could command both cash purses and appearance money. Soon, drag racing had angels, impresarios, press agents, roadies—all the trappings of the entertainment business—and the expression "high performance" had taken on a dual meaning, technological and theatrical.

Forty years from drag racing's beginnings, Orange County Airport would long since have turned into John Wayne International, and not one of the major racing venues would be a rural airstrip or anything like it. Rather, there would be supertracks like Gainesville Raceway in Florida and the Texas Motorplex near Dallas, replete with electronic scoreboards and comfy viewing suites for VIPs. The activity itself would have been transformed into a lavish spectacle with full-blown television coverage and multimillion-dollar gates at a nationwide series of four- and five-day events. Drag racing would have spawned superstars like Garlits and Shirley Muldowney as well as sages like Pappy Hart and Wally Parks, the man who founded the National Hot Rod Association (NHRA) in 1951 as "a semi-social car club," and then built it into a position of drag-racing dominance that was nearly absolute.

To read NHRA's *National Dragster* week after week, year after year, is to bathe in a saga of unilinear progress: ever better performances, bigger crowds, more money. Supported by commercial sponsors, the top racers haul their dragsters around in 18-wheel rigs along with a closet full of spare 4,000 hp. engines. A bare chassis, a sinuous 25 feet of chrome-moly tubing, is worth about $40,000 a pop, but the mobile transporter may entail an investment ten times larger. In a lounge adjacent to the work area, corporate types can munch catered tasties in air-conditioned comfort while crewchiefs pore over computer printouts next door and consult with resident engineers. Some racers have come away from the final event in NHRA's world championship series with nearly a quarter of a million dollars.

Considering that it all began with prewar Fords stripped to bare frame rails (hence the sobriquet "rail job" that still persists), with trophies worth $7.50, and with a strict reverence for seat-of-the-pants pragmatism, this is a compelling example of how far a technological enthusiasm can carry when it gains momentum. Men like Pappy Hart and Wally Parks, both in their 80s, are prone to muse about never having dreamed that "it would get so big," yet there is also a strong tendency to assume an inevitability, to see big-bucks drag racing and high-tech dragsters as "evolving" naturally, much as people tend to assume that ocean liners evolved from dugout canoes, moonshots from skyrockets.

It is easy to disprove this misconception. All one has to do is drive over Cajon Pass and out to Adelanto some summer Sunday morning, thence down a bumpy, one-lane road to El Mirage. There, one will find that hot rodding's other offshoot has taken quite a different course. This is true in a literal sense—a long unpaved course, not a short stretch of asphalt or concrete saturated with chemicals for maximum traction. But the more significant truth concerns motivation. At the lakes, enthusiasm is virtually the sole driving force: "The lakes racers are still doing it for kicks," writes journalist Dave Wallace. There is no commercial signage, no prize

Shirley Muldowney in action at Pomona in 1987.

money, and it costs nothing to get in the gate (only a figurative gate in any event, for there is no fence).

As at the drags, there is a wealth of mechanical ingenuity: speeds well above 200 m.p.h. have been clocked by Ford roadsters and coupes from the 1930s, by motorcycles, and by "lakesters" with bodies fashioned from aircraft drop-tanks, not to mention late-model Detroit iron that looks perfectly stock. Because the timing traps are only a little more than a mile downcourse, El Mirage would be nowhere to go after the ultimate, the land speed record; despite occasional forays elsewhere, the place to seek it is still Bonneville. Even at Bonneville,

At El Mirage in the summer of 1990, Roy Creel and Terry Curian, in driver's firesuit, remove the hood from their 1934 Ford coupe after Terry's return from a run.

This configuration was conventional for dragsters from the latter 1950s until the early 1970s. Said one bystander when he saw the first machine designed like this, "The driver sits back there like a rock in a slingshot," and the name "slingshot" caught on. Here, in Arizona in 1959, Don Garlits has an early lead over Californian Chuck Gireth.

however, one sees scarcely a hint of the huckstering that pervades every major drag race. No inflated Budweiser cans or cigarette packs placed to command a very photographic vista, no commercials coming over the public-address system—indeed, the p.a. is so low-tech that the announcer can scarcely be heard at all. And there are no purses; in the words of a writer for the *Salt Lake Tribune* (only the Utah papers pay any sustained attention to what happens at Bonneville), the racers "scrimp, save and sacrifice in an effort to gain their own little piece of fame."

It costs a few dollars to get in, but spectators must provide their own seating. Here lies the fundamental difference with drag racing: land speed racing has never truly become a spectator sport. At an NHRA drag race, tens of thousands of people come through the gate and head for the grandstands. At Bonneville, or especially the lakes, competitors may well be in the majority. Yes, it would challenge even the sharpest promoter to induce throngs out into the desert to watch "timing," no mat-

ter how fast the cars were going. Perhaps it could be done, but this would entail restructuring the activity with a good measure of theater, as drag racing has been restructured. Few competitors want any part of it. For most of them, the technological challenges are quite sufficient.

Breaking Records

Rough-hewn though they may have been, it seems probable that some of the very first dragsters in the late 1950s could have gathered speed as quickly as any automobile that was ever built. The standing-mile record, held by a German Grand Prix machine, had stood at about 200 m.p.h. since 1937. By 1964 dragsters were routinely clocking 200 in a quarter mile. Before the hot rodders, nobody ever thought much about building a car for the *express* purpose of acceleration, but things moved fast when they did.

Enthusiasm for top speed, on the other hand, went back to the turn of the century—100 m.p.h. was exceeded by 1905 and 200 by 1927. Both marks were set on the

sands of Daytona Beach by Americans, albeit in European machinery. For a long time after that, however, the game was mostly played by wealthy Englishmen. In the 1930s the record was hotly contested by Malcolm Campbell, George Eyston, and John Cobb, driving costly leviathans powered by British aeroplane engines, Rolls Royce or Napier Lion V-12s. Four years after Campbell broke 300 m.p.h. at Bonneville in 1935, Cobb went 367, and in 1947 he returned to the salt and clocked 394. His vehicle was the very antithesis of the hot rodder's ideal of spare simplicity, but there was no disputing the reality that it was more than twice as fast as the fastest machine at the SCTA's first Bonneville Speed Week in 1949.

A decade later there was a flurry of interest in the land speed record. Seeking the LSR, six men ran better than 400 on the salt in the 1960s, and a seventh, Malcolm Campbell's son Donald, did so in Australia. His machine had a 4,100 hp. turbine such as powered the Britannia airliner; three others, Americans, had turbojet engines, and another American had rocket power. More recently a jet machine driven by an Englishman went 633 and a rocket broke Mach 1—the site was Edwards Air Force Base, and one of those who attested to its speed was Chuck Yeager. Such vehicles were essentially aeroplanes without wings, and nobody would have deemed it impossible that one could go over 600, or even the 739 credited to the rocket at Edwards, with its reported 48,000 hp.

There were jet and rocket-powered dragsters as well as LSR machines, but purists in both realms felt that these transgressed some fundamental canon. The purist LSR camp was dominated by American hot rodders whose dream was to find out what was possible using souped-up automobile engines. Bonneville pioneer Mickey Thompson went 406 with Pontiac power in 1960, but never managed a proper run back up the course as stipulated in the rules for international records. For a generation of hot rodders, the ultimate heroes were the Summers brothers, Bob and Bill, from Arcadia, California. In November 1965, they hit a two-way average of 409.277 with Chrysler power, and that mark stood for more than a quarter-century.

During this period, speeds for the fastest dragsters increased from around 200 m.p.h to nearly 300 and quarter-mile elapsed times came down from 7½ seconds to less than 5. This might suggest that financial inducements were crucial. But comparisons between what happened in the two realms can be deceptive. In the 1960s, the Summers brothers used not just one engine, but four of them si-

multaneously, as did Mickey Thompson. In the 1980s other enthusiasts came to the fore whose passion was to find out what could be done with only one engine. Nolan and Rick White, father and son from San Diego, preferred an aluminum Chevrolet; Elwin "Al" Teague of Santa Fe Springs favored a Chrysler. Neither would even consider jet or rocket power. If jets and rockets really counted, said Nolan White, "the government would hold the record. This is really the last racing battleground for the little guy."

In 1990 both White and Teague topped 400, and in 1991 Teague finally broke the Summers brothers' mark, by .6 m.p.h. And he did it, as *Hot Rod* magazine's Gray Baskerville put it, "on his own nickel." "No 18-wheel support ve-

hicles or a bin full of parts," Baskerville continued. "What suffices is the leftovers from a millwright's paycheck, the space afforded by a completely cluttered 1½ car garage, a parts bin full of IOU's, and the support of a battered blue Chevy pickup towing a rusty open trailer."

Dragster racers who set records these days not only have cushy sponsorships but are usually wealthy in their own right. Not that they do not feel as much technological exhilaration as anybody, but having the money for "a bin full of parts" can ease a lot of burdens. Hence what people like Al Teague may accomplish is all the more eloquent testimony to the vitality of their dreams.

Yet their attainments are not the ultimate confirmation of what a

The *Shadoff Spl.,* one of the early Bonneville streamliners built by American hot rodders, is seen on the salt in 1954. Driver Mal Hooper's fastest time was 236.

Mickey Thompson prepares for a dawn record run in his four-engine *Challenger*. Wife Judy stands by with his helmet while a crewman cleans salt from the soles of his shoes.

powerful inducement the "is it possible?" question can be. Beyond the realm of land speed records or the top quarter-mile records, there is a whole array of other marks to shoot for. All of them are slower, but of course it is not some *ultimate* mark that really matters at all. If that were so, the government would indeed hold the record; a space shuttle orbits the earth at more than 17,000 m.p.h. What counts is the speed that can be attained given a set of acknowledged tradeoffs.

A dragster has only 1,320 feet to go, but it can have gobs of horsepower, and apply it to a very sticky surface through huge tires. A Bonneville streamliner has miles to get up to speed, but the driver needs a light touch and the engine must be far milder than a dragster's, which, after all, is only under power for a few seconds. Or forget Bonneville's expanse of glistening salt, drag racing's high-tech paving, and consider what may be possible on the dirt at El Mirage. There was talk in the 1950s about

abandoning this site because it was not suitable for speeds approaching 200; the surface is no better today, but the El Mirage "200 MPH Club" has dozens of members and one of them has actually gone 300 there. Or take yet another form of straightaway contest, one not yet mentioned, the tractor pull—consider what can be done on loose dirt with a ponderous weight in tow.

Which is by way of reiterating that the question "is it possible?" is always posed within a context of personal choices, choices that may be socially constructed but invariably reflect private dreams. Some people are loyal to a particular brand of engine. Some prefer a potent if tricky blend of methanol and nitromethane for fuel; others are partisans of pump gasoline, choosing to call the name of the game mechanics, not "chemistry." By juggling combinations of restrictions on displacement, fuel, and configuration, the range of options is endless.

Every form of straightaway

competition has a "class" structure. At the drags or at Bonneville there may be dozens of winners, and across the board there are literally hundreds of certified world's records. All such records are regarded respectfully, even (or, rather, especially) the slower ones, for there is no mistaking the ingenuity displayed by vehicles that are designed to go fast but cannot possibly go as fast as the fastest vehicles go. Some cynics say that unlimited records are simply a matter of money, but everyone knows that records attained under stringent handicaps require something more precious.

Clearly, many of the most elegantly designed vehicles have resulted from confronting severe technical restrictions. Here one begins to approach the ultimate by leaving the ranks of hot rodders and considering conveyances that are not permitted to have any mechanical power at all, such as Human Powered Vehicles—streamlined, featherlight bicycles. Nobody would think of denigrating HPVs because even the best go less than 70 m.p.h. Or consider the soap box derby racer, which is timed at the end of a downhill run, no application of any power being permitted. The ultimate constraints are frictional and aerodynamic drag, the top speeds about 30 m.p.h. But, other than his or her age, a soap box derby enthusiast is certain to be somebody very much like Al Teague or Don Garlits.

In telling "Here's How I Win Drag Races" in a 1965 issue of *Popular Mechanics*, Garlits estimated that most people involved with dragsters like his earned little beyond "food and fuel" and only a

handful made "a good living." Despite appearances, this is still pretty much true today. Lots of people around drag racing are well off but most of the racers pursue their quest for motives that have little to do with making "a good living." In the other racing constellations absolutely nobody expects to make a living.

In the arcane realms of speed, the question "is it possible?" has yielded awesome performances. Captain Laughlin McKay once drove his brother Donald's *Sovereign of the Seas* on a South Pacific reach of 1,478 miles in four days. He might not have had a total rapport with "Capt. Jack" McClure, designer and driver of a go-cart capable of a quarter-mile e.t. of 6.3 seconds. But they would have understood one another pretty well, for each had a dream and each "held the record."

For the rest of us, however, there remains another question, "So what?" What is the good of trying to understand technological enthusiasm?

The history of technology was once dominated by an orthodoxy called "internalism," which held that there could be no proper analysis without the same mastery of technical knowledge that technicians have. But this was never the real necessity. Far more important is an appreciation of how exciting technological pursuits may

Top: Elwin "Al" Teague on his way to 400 m.p.h. in 1990. *Above:* An ESPN stringer was on hand to interview Teague, left, after his record-breaking run in 1991, but otherwise the scene is decidedly low-key.

be to their practitioners—*all* technological pursuits, not just essentially benign varieties such as described here. On the dark side, a sense of exhilaration is likewise what impels the quest for monsters of domination, exploitation, and destruction. "Whether or not historians of technology themselves feel this exhilaration," Ferguson writes, "we must still understand and appreciate it if we are to understand and appreciate the unfolding of America's technological history."

Technological dreams can never be fully explained in economic terms. Brooke Hindle suggests that men like Oliver Evans and William Norris were driven by a need to fulfill new potentials simply because those potentials existed. He suggests further that technology may harbor some sort of "elemental force." Positing the existence of such a force is not the same as saying that technology ever "determines" anything, but it surely does say a lot about how it gains momentum—how *any* technology gains momentum. This would seem to include some exceedingly costly technologies that are rationalized on the basis of concerns like "national interest," to be financed by taxpayers, when in fact the fundamental push may be nothing more than the human enthusiasm that an unfulfilled potential is bound to kindle.

FEBRUARY 1949

35 CENTS

POPULAR MECHANICS
MAGAZINE
WRITTEN SO YOU CAN UNDERSTAND

BUILD YOUR OWN URANIUM DETECTOR - *Page 238*

Although most people were permitted nowhere near any aspect of nuclear technology in the early postwar years, the image of the uranium prospector preserved the illusion that ordinary citizens could participate in the atomic adventure.

"Planetary Engineering": The Strange Career of Progress in Nuclear America

HOW DID AMERICANS THINK ABOUT

TECHNOLOGY IN THE EARLY POSTWAR

DECADES? ALTHOUGH WE CANNOT

ANSWER THIS QUESTION DIRECTLY,

WE CAN LOOK FOR CLUES IN THE

THINGS THEY SAW AND HEARD.

BY EXAMINING REPRESENTATIONS OF

TECHNOLOGY AT MIDCENTURY, WE MAY

DISCERN PATTERNS THAT HELP EXPLAIN

WHAT HAS CHANGED SINCE THOSE

YEARS OF GREAT PROMISE AND

COLLECTIVE SELF-DECEPTION.

MICHAEL L. SMITH

"The Miracle of U-235." "Oxydol—With Tiny Green Crystals!" "Space: The Final Frontier." "'56 Dodge—The Magic Touch of Tomorrow!" "Without Chemistry, Life Itself Would Be Impossible." "Progress Is Our Most Important Product." From the daily headlines to the commercials on their new televisions, in countless government pamphlets and corporate publicity campaigns, in comic books, schoolchildren's copies of *My Weekly Reader,* and Boy Scout merit badges—even in speeches on the floor of Congress—wherever they turned, post–World War II Americans were reminded of the wonders of technology. More than that—they were assured that the future well-being of their families, and of their nation, was hitched to technology's ever-rising star. Americans became so accustomed to measuring their national and personal progress in terms of new Buicks and space capsules and nuclear submarines that the twin assumptions of American technological optimism were seen as inevitable axioms: that progress is irreversible, and that technology is its delivery system.

This elevation of technology to signify national identity and progress did not originate in the 1950s. Ever since colonial revolutionaries memorized British designs for manufacturing rifles with interlocking parts (and later reproduced them), Americans have attached their collective fate to machines. Historians have tried to explain this fondness for techno-visions. With so much less history than other industrializing nations (according to conventional wisdom), Americans looked forward rather than to the past, and to technology rather than to other areas of cultural achievement for the key to their impending greatness. Words like "progress" and "technology," however, have changed in meaning over time, conjuring up different artifacts and social processes for different generations of Americans.

Before World War II, the United States was one rising industrial power among many; "progress" meant that the standard of living appeared to be improving for the majority of citizens, and that the nation's importance in the world was approaching that of other global leaders. Not until the Great Depression of the 1930s did Americans hear widespread concerns that these measures of national greatness might not continue to spiral upward unabated after all.

World War II changed everything. By 1945, wartime production had ended the Depression; the United States was the only major power that had not suffered extensive homefront damage; and American leaders possessed an "atomic monopoly." As Winston Churchill observed after Hiroshima, "America stands at this moment at the summit of the world." But what direction does one choose, having once reached the summit? And what happens to technological optimism when the most significant technological breakthrough of the period is also the most destructive weapon ever devised? The postwar era promised unprecedented affluence and prestige; but that new abundance and power also tested the culture's traditional conflation of technological enthusiasm and faith in national progress—sometimes in unexpected ways.

How did Americans think about technology in those early postwar decades? Although we cannot answer this question directly, we can look for clues in the things they saw and heard. By examining representations of technology at midcentury, we may discern patterns that help explain what has changed since those years of great promise and collective self-deception.

Among the many sources that might help us assemble images and ideas about technology in early postwar America, *Popular Mechanics* is a useful place to start. Virtually all of the assumptions and predictions about technology found in its pages were present in the nation's other mass-market magazines; but *Popular Mechanics* offers a high-density sampling of prevailing attitudes. "Written so you can understand it," as each issue promised, the magazine had struggled since 1902 to bridge the ever-widening gap between the words of experts and the enthusiasm of amateur technophiles.

Fifty years later, individual inventors and do-it-yourselfers were still featured; and each issue continued to offer its (male) readers advice on new projects in the shop or in home improvement. But the magazine's vision of the postwar world was reflected, in part, by the more or less regular columns that appeared for the first time in the 1950s: "Detroit Listening Post," "An Eye on Space," "Keeping Up with the Atom," and "Sidelights from the Pentagon."

As these topics suggest, *PM*, as it had taken to calling itself, tried to make a place for "Big Science" alongside its readers' weekend projects. In spite of the cognitive dissonance that sometimes resulted—the December 1952 issue, for example, featured "Playing Safe with Atomic Rays" back-to-back with "Selecting Your Christmas Tree"—*PM* tried to make space travel and atomic power seem as accessible to its readers as the family car.

But if the automobile permitted "hands-on" familiarity for the consumer, nuclear reactors and lunar probes did not. Nor could either the readers or the editors of *PM* fail to notice the growing distance between individual citizens and technological design and implementation. In the 1950s, national lore continued to celebrate scientific and technological change as the world of individuals— Edison, Lindbergh, Ford, Einstein—even though (or perhaps because) the myth of the individual on the frontiers of science had faded into anachronism. Research and development had long since become the domain of large corporations and, increasingly, of the federal government. In order to allow its individual readers to imagine Big Science on a human scale, *Popular Mechanics* bestowed a

Long before the U.S. Government approved of nonmilitary applications of nuclear technology, atomic optimists envisioned a postwar technological utopia as inviting as the Bomb was terrifying.

kind of "male domesticity" on the large-scale projects it described. The gender cues for science and technology remained masculine; but Big Science became domesticated—reduced, by a series of rhetorical devices, to a scale comparable to that of the basement workshop.

Nowhere was this task more challenging than in the realm of nuclear technology. Conceived in unprecedented secrecy, delivered to the world in the form of a weapon of unimaginable destructiveness, nuclear technology had to be represented in ways that outdistanced that secrecy and destruction—repackaged as a cornucopia of consumer benefits. Americans were being asked to embrace a technology they had never seen, except in photographs of mushroom clouds and devastated cities.

From the standpoint of nuclear publicists, the general public's lack of first-hand familiarity simplified the task of recreating the image of the atom. In the 1950s and early 1960s, nuclear promotion had not yet been challenged by the debates over reactor safety, thermal pollution, waste disposal, and community evacuation procedures that characterized the '70s and '80s. Even the incomplete but growing body of information on the dangers of radioactivity and fallout was systematically suppressed by the Atomic Energy Commission. And the rise of postwar environmentalism, with its questioning of the most basic premises underlying Big Science, was still limited to a few plaintive voices.

During these "golden years" of Atoms for Peace, enthusiasts portrayed nuclear-powered electrical power plants as just the beginning of a new era of "unlimited power." More than a technology, nuclear power became a Rorschach test, revealing the fantasies and apprehensions of postwar political culture. (What, after all, would any of us do if we suddenly acquired "unlimited power"? What would we change? And how would such power change us?)

But Big Science (and nuclear power in particular) was also more than a passive recipient of cul-

Hollywood contractor Hal Hayes was one of many shelter designers who insisted that Americans could survive the atomic holocaust in luxury.

tural meanings; its proponents assigned political and cultural attributes to the technology—and, implicitly, to their national audience. Already a culture of technological optimists, postwar Americans heard a dazzling array of stories from their nuclear Scheherazade—stories that intermingled the fate of their listeners with that of nuclear power.

One way to domesticate nuclear technology was to depict people prospecting for uranium—perhaps the only part of the nuclear production process accessible to ordinary, individual citizens. *Popular Mechanics* ran frequent stories about successful prospectors (the "new 49ers"), and encouraged readers to build their own Geiger counters. ("It sometimes takes no more than enterprise, luck and a little do-it-yourself talent to make a full-fledged uranium king," declared an October 1955 feature about a fireman who hit radioactive paydirt with a Geiger counter made from plans that had

been published by *Popular Mechanics*.)

Civil defense was one of the few aspects of the atomic age in which the government encouraged active citizen participation; and *Popular Mechanics* was quick to incorporate fallout shelters into its pantheon of projects for the household craftsman. As the magazine articles revealed, fallout shelters presented excellent, if eerie, opportunities to suburbanize the bomb and its threat of nuclear holocaust. Here, after all, was a nuclear project within the power of the individual citizen, and it reduced the dimensions of nuclear war to the level of adding a new room to the house.

In August 1953, readers of *Popular Mechanics* learned of a "House for the Atomic Age" built by a Hollywood contractor. Designed to "bring the outdoors indoors"—an odd function for a fallout shelter—the luxury home featured a "swimming pool that becomes an automatic decontamination bath during an A-bomb attack." In March 1958, an article on "Atomic Hideouts" noted that the Gaither Report, jointly issued by the Office of Defense Mobilization and the National Security Council, called for a national system of public fallout shelters. In keeping with its do-it-yourself ethic, *PM* advised readers to build their own personal shelters. The magazine suggested that underground shelters could serve double duty, providing "a cool place in which to relax during the hot summer weather." A December 1961 feature assured readers "You *Can* Build a Low-Cost Shelter *Quickly*," and provided four models, ranging from a $30 plan (consisting of a hundred sandbags) to an elaborate $1,800 "Under the Patio" design suitable for the concerned suburbanite.

Radiation could also be domesticated by describing it in the language of everyday events. One of *PM*'s earliest articles to address the dangers of nuclear technology was "Playing Safe with Atomic Rays," which appeared in December 1952. Radioactive materials, staff writer Rafe Gibbs explained, were like psychotics—potentially dangerous, but manageable: "If not put under control, an unstable man can be dangerous. Same with unstable isotopes. . . . Both humans and isotopes sometimes 'blow their tops.'" With enough control, there was no real danger. The article talked of "Joe," a bus driver who took a job at the government's nuclear facility at Hanford, Washington: "He feels safer than he did driving a bus."

The atom could also be domesticated by associating it with everyday household goods. "Spray Bomb and A-Bomb: Cousins Under the Shell" (*PM*, June 1950) asserted that since "Fluorine was tamed by the scientists who put together the atom bomb," the proliferation of aerosol sprays (from bug-bombs to deodorants) was "a whopping first installment on the atomic energy project's promise of a new world."

A July 1958 article by Clifford B. Hicks depicted a young woman applying make-up while a mushroom cloud rose in the background. "Meet Lithium," the title proclaimed: "The Stuff of H-Bombs—and Milady's Powder." Lithium, a crucial element in nuclear weapons research, was an intimate part of our lives, Hicks argued:

You use it countless times each day without recognizing it. Probably your car is lubricated with lithium grease. The vitamin A you take in the morning is manufactured with the aid of a lithium chemical. You may eat your dinner off a plate glazed with lithium and take your bath in a tub coated with lithium enamel. When you watch TV you're looking into lithium glass.

Lithium is in carbonated drinks, and "Milady uses lithium stearate when she covers her cheeks with face cream." Despite this versatility, Hicks observed, lithium had long been considered "the poor cousin of the other alkali metals, a ne'er-do-well actor who never would amount to much." But now, thanks to its role in "taming the H-bomb," lithium "may turn out to be the rich uncle who turns up unexpectedly to solve the family's problems."

If ascribing nuclear kinship to the contents of

Like the uranium prospector, the do-it-yourself home owner could "domesticate" the atom by constructing a low-cost fallout shelter.

At 11:00 o'clock on the morning of August 12, 1952, the Atomic Energy Commission announced in Washington, D. C., that it was investing $1,000,000,000 of the taxpayers' money in a vast new gaseous-diffusion plant. This atomic factory would be built into the Scioto River Valley of southern Ohio.

A few minutes later a congressman picked up his phone and placed a call to the old industrial town of Portsmouth, Ohio. The call brought the news to Portsmouth. For a decade the town had been slowly fading. Jobs were scarce, school kids disappeared upon graduation, and the population had been dropping steadily since the depression.

By midnight it was apparent to the people of Portsmouth that some sort of atomic bomb had been dropped on their town. The hotels suddenly were jammed, long-distance lines were tied in knots and speculators were swarming into the area.

The next morning trailer-camp operators started dickering for choice tracts along Highway 23 which runs north out of Portsmouth to the area selected for the atomic plant some 22 miles away. Scouts for restaurant chains hustled into town, and speculators were picking up options on cornfields for future drive-in theaters.

The atom had come to the

POPULAR MECHANICS MAY 1956

WRITTEN SO YOU CAN UNDERSTAND IT
VOL. 105   NO. 5

Security force stands eternally vigilant over huge atomic plant. Cloud of vapor rises night and day

Look What Happens When

THE ATOM STRIKES A VALLEY

By Clifford B. Hicks

Post-construction sale brought thousands who bid on items from houses to tools

At the dawn of the nuclear age, the social consequences of nuclear technology were measured in terms of jobs and new economic opportunities. In subsequent decades, a number of "atomic cities" had second thoughts about the benefits of the atom.

bug bombs and cold cream was not sufficient to rid the atom of its intimidating image, writers could portray nuclear processes themselves through household analogies. In January 1959, "Fusion Power for the World of Tomorrow," another article by Hicks, described how "harnessing the H-bomb for peaceful uses" would produce "billions of years" of fuel through a process "somewhat similar to the burning of logs in your fireplace." When fusion occurs, Hicks explained, atomic particles collide "like ants on a hot stove." To achieve this effect, "the scientist must create a star on earth, bottle it up and clamp on a lid. If he succeeds, he can draw from the bottle any amount of power he wishes."

As the star-in-a-bottle image implies, the domesticated atom was by no means a disempowered one. On the contrary, two apparently contradictory strategies characterized popular images of Big Science, in *Popular Mechanics* and in mass culture generally: in addition to the images of domestication (of which we have been exploring one variety), depictions of the future of technological America conjured scenes of *colonization*. By a kind of geophysical imperialism, proponents of Big Science envisioned a planet as subservient to its mas-

Scientists at Lawrence Livermore Laboratory envisioned instant harbors and canals through "Project Plowshare," a series of earth-moving proposals utilizing thermonuclear explosives.

ters' wishes as any conquered military opponent. Before we can conjecture about the relation of domesticating and colonizing images, we need to examine the latter as they were presented.

In *PM*'s October 1958 issue, Harland Manchester described "The New Age of 'Atomic Crops.'" At the Atomic Energy Commission's laboratory at Brookhaven, he wrote, scientists had developed the "world's first radioactive farm" in an effort to "speed up and control the slow, erratic processes of natural evolution by means of the new tools of the atomic age." Manchester quoted a Swedish geneticist's opinion that "Most food plants are rather old-fashioned, and their variations have been exhausted by inbreeding. They need to be reconstructed to suit the needs of modern agriculture, with its emphasis on high yield and mechanization." With nuclear technology, he concluded, "We now have an instrument with which we can rebuild all the food plants in the world."

As far-reaching as it sounded, the "restructuring" of plants paled before other proposed applications of "unlimited power." One of the most ambitious projects of nuclear enthusiasts was Project Plowshare, a series of proposals for canal and harbor construction, mining, and other earth-moving tasks—with thermonuclear bombs. Plowshare originated during the Suez Canal crisis of

1956, when the threat of a closed canal prompted a group of government scientists at the Lawrence Radiation Laboratory in Livermore, California, to propose the excavation of a new canal (perhaps through Israel) with nuclear explosives.

Although that project never materialized, this group of scientists proposed and designed a broad variety of nonmilitary tasks for nuclear bombs over the next 15 years: "mining" metals, oil, and natural gas with nuclear detonations; a passageway through the mountains in southern California for a highway and rail line; an "instant harbor" on the north shore of Alaska; a "Panatomic Canal" through Nicaragua. Most of Plowshare's proposals were never implemented; only a few made it to the preliminary testing stage. But for nuclear publicists, Plowshare provided both a challenge and an opportunity.

One of Project Plowshare's most prominent publicists was Edward Teller. A brilliant physicist, "Father of the H-Bomb," and a fierce Cold Warrior, Teller was also a tireless publicist for nuclear technology. A founding director of the Lawrence Livermore labs in the 1950s, he was among the first, and the most emphatic, Plowshare proponents. "Nuclear Miracles Will Make Us Rich," proclaimed a February 1959 article by Teller for *This Week* magazine, a syndicated insert for Sunday

Michael L. Smith

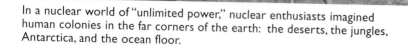

In a nuclear world of "unlimited power," nuclear enthusiasts imagined
human colonies in the far corners of the earth: the deserts, the jungles,
Antarctica, and the ocean floor.

newspapers. (He contributed a similar piece to the March 1960 issue of *Popular Mechanics.*)

Teller assured his readers that Plowshare would usher in "a decisive *victory* in man's historic *battle* to shape the world to his needs." Excavation with H-bombs could "dig harbors and canals that will open vast areas of the world to commerce and development," but that was only the beginning. With this vast new power at our command, the earth and its natural resources could be fundamentally rearranged: "We can *free* immense oil and mineral reserves from their *rocky prisons*;" "*blast* the earth cover from extensive ore deposits so we can *scoop up* the minerals"; "*squeeze* oil from rock"; "*blast* new courses for underground rivers" and "convert *desert wastes* into green fields" with water from rivers that, unless diverted, "flows *uselessly* to the sea." His choice of words demonstrates how, in the deeply sublimated language of the Cold War, images of conquest and colonization could cast nature as a stand-in for the adversary.

According to Teller, the sky as well as the earth awaited reconfiguration by the atom: "Outlandish as it sounds, we may even be able to *control our weather*" by using nuclear explosives as a "trigger" to "decide where rain shall fall and where it shall not fall." Plowshare, he wrote, "is on such a vast scale that I would call it *geographic engineering.*"

Another prominent celebrant of the atom was Glenn Seaborg, a co-discoverer of plutonium who directed the Atomic Energy Commission from 1961 to 1971. Although very different from Teller in his outlook on nuclear weapons (he encouraged arms talks while Teller promoted new weapons), Seaborg shared his colleague's enthusiasm for Plowshare. In countless press releases and articles, in comments for newspapers and magazines and in *Man and Atom*, a book he co-authored, Seaborg urged the public to support the unfolding adventure of "building a new world through nuclear technology."

While Teller advocated "geographic engineering," Seaborg called for "planetary engineering" to correct and improve "a slightly flawed planet." "New worlds above and below" awaited us, he wrote: "The oceans and the Earth's rocky mantle remain almost untouched. With the application of

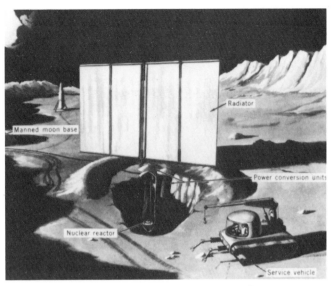

Glenn Seaborg, director of the Atomic Energy Commission, assumed that in the near future nuclear power would generate lunar mining colonies.

abundant nuclear energy even these frontiers will open." In Antarctica, the small nuclear power plant at McMurdo Sound could serve as a "prototype" for self-sustained underground colonies throughout the Earth's "nether frontier." On the moon, nuclear power plants would permit "lunar mining and manufacturing" by colonists. Eventually, interplanetary nuclear rockets and extraterrestrial nuclear excavation might lead to colonies flourishing inside hollowed-out asteroids.

Unlike most nuclear enthusiasts, Seaborg acknowledged in 1971 that part of the appeal of these colonies was the escape they could provide from "the polluted ecosphere that society is now brewing," as well as from "man-made catastrophes—as man's last retreat in the event of total nuclear war." He observed, "On the moon, at least we have the opportunity to begin afresh and, we hope, more wisely."

Seaborg preferred to describe nuclear-powered colonization, however, as a precondition for the birth of "three-dimensional life." "Mankind's normal habitat is a thin, essentially two-dimensional layer clinging to hospitable areas on the planet's surface," he wrote; but the "atom-based tools" of the near future "will help us *cut the umbilical cord* that ties us precariously to the thin layer of air, water, and rock coating this 8000-mile sphere ca-

Visitors to the General Motors "Futurama II" exhibit at the 1964–65 New York World's Fair witnessed depictions of the imminent "conquest" of earth, sea, and space in the nuclear age.

reening through space. Technology," he concluded, "gives us the power to become three-dimensional."

The components of Seaborg's 3-D vision and Teller's geographic engineering—colonies in space, on the ocean floor, underground, in the jungle, the desert, Antarctica—were already familiar to readers of *Popular Mechanics* primarily as projected manifestations of limitless nuclear power. In the 1950s, the "conquest" and colonization of Antarctica became a popular theme, resurrecting the frontier images of a previous era. In "Tracks Across the Polar Continent" *Popular Mechanics* (June 1961) Richard F. Dempewolff, in one of a series of articles on Antarctica, characterized "the men who are out there probing the white wilderness" as "foot soldiers in a war" who were "pulling back the veil from a continent that was all mystery just a few years ago."

Colonization of the rainforests was also a popular theme. "Jungle Road Knifes through the Last Frontier," proclaimed a September 1959 feature by Thomas E. Stimson, Jr. "The world's greatest jungle, the 'green hell' of the Amazon that has withstood civilization for centuries, is finally being conquered," Stimson wrote. The agent of civilization, in this instance, was "a 1400-mile highway [that would] pierce the Amazon heartland."

Technological advances in the war on vegetation won enthusiastic plaudits. In an article titled "He's Changing the Face of the Earth," Paul Lee profiled the developer of "a new breed of mechanical monsters—land-clearing machines—designed to clean up the face of the earth" by uprooting and pulverizing trees in a few deft strokes. The need for such equipment seemed clear:

> In this country there are 50 or 60 million acres of fertile land covered by unproductive trees and brush. And more important, an umbrella of rain forest covers billions of acres of potentially good farm land in Africa and South America.

And when the South Pole, the Amazon, the deserts, and the other remaining corners of uncolonized terrain had been "tamed," there remained two more frontiers: the oceans, and outer space. Cover stories featured "conquest of the deep" with giant atomic sub tankers, "bathyscaphs," and experimental flying-wing shaped "gliders." After the launch of Sputnik I in October 1957, the space race provided ample opportunity for *PM* to domesticate outer space through such articles as "It's Lonesome Up in Space" and "I Lived with the Astronauts."

But space also served as the ultimate arena for visions of future colonization. An August 1952 article by staff writer Stimson asked, "Shall We Move to Another Planet?" Based on an interview with Caltech astrophysicist Fritz Zwicky, Stimson's piece announced that "the time is coming when mankind will have to reconstruct the solar system," although "we may have to rearrange the planets and in some cases rebuild them to fulfill our future requirements for living space." Before colonizing it, we might need to "send Mars off on a tangent past some other planet to draw off an atmosphere"; the outer planets would have to be "broken apart or shrunk" and then "moved closer to the sun," the article proclaimed.

All of this would be only part of the human use of the "unlimited power [that] will be available when nuclear fusion is achieved," providing "the tool with which the planetary system could be juggled about." In an analogy that would become familiar to space enthusiasts, Stimson quoted an Air Force space medicine researcher's observation that "the conquest of the outskirts of the atmosphere and eventually space is a revolutionary event comparable only to the transition of aquatic animals to the land in geological times."

These visions of universal human conquest and colonization of nature were by no means unique to *Popular Mechanics;* they permeated mass culture throughout the 1950s and early 1960s. Visitors to the 1964–65 New York World's Fair found that the most popular exhibit, the "Futurama II" ride at the General Motors pavilion, featured dioramas of human colonization of the moon, Antarctica, the ocean floor, a cultivated desert, and a defoliated Amazon—all accomplished primarily through the application of nuclear power.

Earlier I suggested that we need to reconcile these visions of colonization with their apparent opposites, the images of nuclear domestication with which we began. Now I would like to propose that we consider them as elaborations of each other. At first glance, rechanneled rivers, reconfigured mountains, transformed deserts growing designer crops, colonies on the ocean floor and the planets might seem far beyond the familiar dimensions of everyday life. Viewed closely, however, these visions of colonization can also be recognized as exercises in domestication. The imagined colonies suburbanized the desert, the ocean floor, Antarctica, and the moon; in so doing, they gave value to the massive projects of Big Science. Domestication and colonization were part of the same process.

Beneath the apparent incompatibility of these two depictive strategies lies an array of contradictions imbedded in the relation of militarism to postwar culture. The United States entered World War II, and the many subsequent battles of the Cold War, declaring that it was different. Other nations might act purely to maximize their power; but America stood for something more than that. Democracy, equality of opportunity, the right to self-determination—these were the values that Americans wanted to bring to the rest of the world.

America's acquisition of the Bomb marked a bifurcation between the nation's citizens and its leaders. Both spoke the rhetoric of America's sense of mission; but "realists" within the emerging national security state reasoned that in its newly empowered role, the United States must anticipate and eradicate every effort to challenge its new supremacy. Like the Bomb before it, nuclear-age security became a matter of utmost secrecy; and in its effort to second-guess the adversary, the American government took on the very qualities it purported to fight against. Installing petty dictators around the globe, American leaders derided the battlecries of insurgent movements—democracy, equality of opportunity, the right to self-determination—as the work of Communists. At home, the nation's government and business leaders had the most prosperous period in American history at their disposal; the prime beneficiaries of that affluence were the already-wealthy and the national security state. Utopian predictions of the abolition of hunger and injustice gave way to new weapons for new wars.

The resulting culture was one filled with contradictions. While undertaking aggressive military build-ups, the nation's leaders denounced the ascendancy of militarism in other nations. While committing billions of dollars to the arms race and the space race, American leaders denied that geopolitical competition was a motivating factor.

Even within the discourse of weapons procurement, Presidents and legislators had to speak in self-contradictory terms. Since nuclear weapons were presumably too terrible to detonate, their real power resided in the image they conveyed. To the nation's adversaries, American leaders communicated a confidence in their nuclear superiority, and a willingness to utilize it. To American taxpayers, however, the message was inverted: America would never use its nuclear arsenal, unless forced to do so; but the pitiful state of our defenses rendered us dangerously vulnerable unless new funds were forthcoming at once. In an age when credibility was everything, images were the only real weapons; and the distinction between the protected and the targeted became increasingly blurred.

It is important to recognize that the history of *images* of technology (and of nuclear technology in particular) exists on its own terms, often quite apart from the history of *applications* of that same technology. Whether the colonization of the moon or the rainforest was *ever* likely is of little significance in the realm of image management; rather, what matters is whether prevailing images of large-scale technological projects helped to convince American citizens, and assure their government and corporate leaders, that the projects they *did* undertake were justified and wise.

No conspiracy theory is required to understand this manipulation of images of technology; for in a culture so steeped in contradictions, even those in power need to subscribe to a vision that addresses those contradictions—if only by burying or disguising them even as it acknowledges them. Thus the colonies—on the moon, on the ocean floor, at the South Pole—signify fallout shelters, even if that function is never articulated.

Nor does one have to venture far to encounter evidence that early postwar America's visions of colonization have been curtailed or discarded. Two editorials that appeared in the June 6, 1991, issue of the *New York Times* are a case in point. The first, "Space Yes; Space Station No," argued that despite $5.6 billion in preliminary spending, NASA's proposed space station should be canceled. It would not further the human uses of outer space, stressed the *Times*, and it would be foolish to "squander precious resources on an empty, costly symbol." Just below that editorial, a second one asked, "What's the Hurry in Antarctica?" After the 26 signatories of the Antarctic Treaty had agreed to prohibit mining there for 50 years, the Bush Administration's State and Interior departments were threatening to oppose ratification on grounds that it would lock up potential resources for too long. The *Times* argued that Antarctica's integrity had global significance, since oil spills and industrial activity could kill photoplankton and warm the ice cap, decreasing the Antarctic region's capacity to convert global carbon dioxide to oxygen.

Consequently, the "decision to invade Antarctica ought never to be made lightly or quickly."

Other aspects of the colonization vision also appear to have been severely discredited. During 1991, a consortium of governments, international banks, and environmental organizations sought ways to forgive portions of Brazil's debt in exchange for that country's agreement *not* to clear so many acres of the Amazon; meanwhile, children at every shopping mall in America were sporting "Save the Rainforest" T-shirts. And if the "T-shirt test" is a reliable index of popular culture, the preservation of whales, dolphins, and their threatened undersea environment has caused widespread concern over the pollution of the ocean by *land-dwelling* humans; the popularity of human colonies on the ocean floor would appear to be waning. (The only T-shirt I have seen that refers to lunar colonies is one that said, "If We Can Put a Man on the Moon—Why Not All of Them?")

How have the expectations of Americans changed? They no longer expect nuclear or any other technology to deliver unlimited power; and they are far more aware of the consequences, social and environmental, of the use of such power. If they no longer believe Big Science is capable of cultivating the deserts or colonizing the planets, they also no longer wish to see the jungles cleared, the oceans mined, or Antarctica stripped of natural resources—even if such projects were feasible. Perhaps most important, Americans in the post–Cold War era may be more difficult to convince that the largest arsenal of weapons in human history produces—or even permits—security.

After so much consideration of past images of the future, we should be entitled to our own techno-visions. Here is mine: the desire of *Popular Mechanics* to make technology accessible has been laudable. Perhaps the time has arrived when decisions about the social uses of technology should finally conform to the dimensions of the household and the classroom—not from images of domestication, but through a genuine democratization of the decision-making process that is responsible for every Dodge, every Stealth, every unheated flat, every mother denied prenatal care in the most powerful nation on earth.

ANTI-GREENHOUSE TECH

BY ABE DANE, Science/Technology Editor

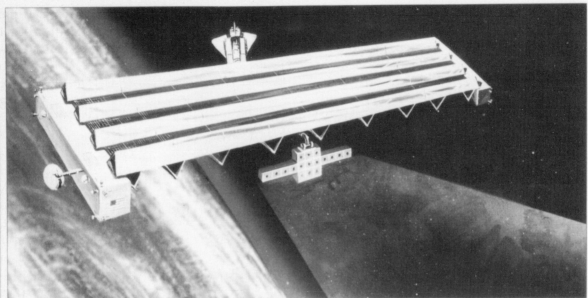

● What if the greenhouse effect is worse than we think? What if we've already pumped enough carbon dioxide into the atmosphere to bring a runaway heat wave that will flood coastal cities, wither crops and wreck our economy?

Although unlikely, such a scenario is within the range of possibilities presented by the latest climate data. So far, supercomputer models have determined only that there is global warming in our future. How bad it will be is anyone's guess. But if worst-case projections are true, simply cutting output of greenhouse-inducing pollutants won't be enough. The only way out of the greenhouse will be the way we got in—through technology.

So far, methods of intervening directly in the mechanisms of global climate have received limited, and usually skeptical, attention from scientists. But an as yet unreleased study by the National Academy of Sciences is reported to weigh several such options and to judge some of them worth further consideration.

The most radical concepts are satellites that would reduce the amount of solar radiation entering our atmosphere. According to Lyle M. Jenkins of NASA's New Initiatives Office at Johnson Space Flight Center, space offers the unique advantage of access to the global climate system as a whole. "You're able to view and interact with very large areas," he says.

The simplest approach would be to erect a giant shade between the Sun and Earth, where it would act as the Moon does during a solar eclipse. Of course it would be on a smaller scale, requiring that roughly 100 sq. km of lightweight material such as mylar be unfolded to have a cooling effect.

An alternative concept would use space simply as a location for energy production, providing a substitute for CO_2-producing powerplants on Earth. Although the idea of using orbiting photovoltaic cells for electricity that could be beamed to Earth is not new, it is getting new attention as photovoltaics grow more economical.

Other orbiting equipment (shown above) could help save the stratospheric ozone layer. Now threatened by chlorine from chlorofluorocarbons used in air conditioners and spray cans, the ozone shield prevents cancer-causing ultraviolet rays from reaching Earth's surface. Theoretically, aiming a solar-powered microwave or charged-particle generator into the upper atmosphere could prevent ozone destruction by neutralizing chlorine atoms.

The main problem right now, according to National Academy of Sciences sources, is that the costs of getting the necessary equipment into orbit are prohibitive. But next-generation spacecraft based on the Advanced Launch Development Program or the National Aerospace Plane could bring those costs down by an order of magnitude. For now, Jenkins believes it's important to gather hard data on how space technology could help the environment. "There's knee-jerk response to messing around with Mother Nature," he says. "But the fact is that we've been messing around with it for a century with our use of fossil fuels." **PM**

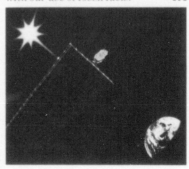

The greenhouse is forcing a new look at solar-powered satellites, such as this NASA concept.

One of the drawbacks of postwar technological optimism was that "technological fixes" often were deployed without regard to their environmental consequences. For latter-day true believers, even the most dire warnings simply called for yet another technological fix.

Further Readings

John L. Wright
Introduction

As mentioned in the text, Thomas Hughes's fine book, *American Genesis: A Century of Invention and Technological Enthusiasm 1870-1970* (New York: Viking, 1989), has served as a source and inspiration for this present volume. Hughes's own collection of historical pieces, *Changing Attitudes Toward American Technology* (New York: Harper & Row, 1975) is also a valuable introduction to the subject. One of the first, and still one of the best, treatments of the popular and literary responses is *The Machine in the Garden: Technology and the Pastoral Ideal in America* (New York: Oxford University Press, 1964) by Leo Marx. An even earlier and more comprehensive study is Siegfried Giedion's *Mechanization Takes Command* (New York: Oxford University Press, 1948).

Civilizing the Machine: Technology and Republican Values in America, 1776-1900 (New York: Penguin Books, 1977), by John F. Kasson, is most pertinent and important to the topic. Other relevant books include: Otto Mayr and Robert C. Post, eds., *Yankee Enterprise: The Rise of the American System of Manufactures* (Washington, D.C.: Smithsonian Press, 1982); David A. Hounshell, *From the American System to Mass Production: The Development of Manufacturing Technology in the United States* (Baltimore: The Johns Hopkins University Press, 1984); and David F. Noble, *America by Design: Science, Technology, and the Rise of Corporate Capitalism* (New York: Knopf, 1977).

The many essays and books of Lewis Mumford offer an informed critical perspective on the subject, as do the issues of *Technology and Culture,* the journal of The Society for the History of Technology.

Joseph J. Corn
Educating the Enthusiast

Fred H. Colvin, *60 Years with Men and Machines* (New York: McGraw-Hill, 1947; reprint edition, 1988). Chatty autobiography by a machinist who, as an editor of *American Machinist,* was a successful and pioneering technical journalist. Joseph J. Corn, "*Popular Mechanics,* Mechanical Literacy, and American Culture, 1900–1950," in Robyn Hansen, ed., *Reading in America* (New York: Greenwood Press, 1992). An examination of the how-to-do-it content of America's premier home mechanics magazine. Eugene Ferguson, "The Mind's Eye:

Non-Verbal Thought in Technology," *Science* 197 (26 August 1977): 827–36. Provides excellent historical background on the impact of printing and early printed illustrations on technological innovation. Steven M. Gelber, "A Job You Can't Lose: Work and Hobbies in the Great Depression," *Journal of Social History* 24, 4 (June 1991): 741–66. An interesting interpretation of the hobby boom during the 1930s, relevant to understanding hobbies generally.

Carroll Pursell
Boy Engineering

The literature of the history of masculinity is only now beginning to be written. For good examples see the essays in Mark C. Carnes and Clyde Griffin, eds. *Meanings for Manhood: Construction of Masculinity in Victorian America* (Chicago: University of Chicago Press, 1990) and Mark C. Carnes, *Secret Ritual and Manhood in Victorian America* (New Haven: Yale University Press, 1989).

Useful information on attempts to guide the development of boys may be found in David I. Macleod, *Building Character in the American Boy: The Boy Scouts, YMCA, and Their Forerunners, 1870–1920* (Madison: University of Wisconsin Press, 1983). The gendering of science and technology toys is covered by the author in "Toys, Technology and Sex Roles in America, 1920–1940," *Dynamos and Virgins Revisited: Women and Technological Change in History. An Anthology,* ed. Martha Moore Trescotte (Metuchen, N.J.: The Scarecrow Press, 1979): 252–67.

Susan J. Douglas
Audio Outlaws

The best general histories of broadcasting in the United States are Christopher Sterling and John M. Kittross *Stay Tuned: A Concise History of American Broadcasting* (Belmont, Calif: Wadsworth Publishing Co., 1990) and Erik Barnouw, *A History of Broadcasting in the United States,* 3 vols. (New York: Oxford University Press, 1966, 1968, 1970). The most detailed information on the early amateur operators can be found in Susan J. Douglas, *Inventing American Broadcasting, 1899–1922* (Baltimore: The Johns Hopkins University Press, 1987). General histories of the phonograph include Oliver Read and Walter Welch, *From Tin Foil to Stereo: Evolution of the Phonograph* (Indianapolis: H.W. Sams, 1976) and Roland Gelatt, *The Fabulous Phonograph, 1877–1977* (New York: Macmillan, 1977). The recollections of amateur operators can be found at the Columbia Oral History Library in New York City. Contemporary accounts of the hi-fi craze exist in various issues of *Business Week, The Saturday Review, Time,* and *Newsweek* in the early 1950s. The best accounts of the underground FM movement are in issues of *Broadcasting,* the industry trade journal. Analysis of how music and music criticism have been "gendered" over the years appear in Susan McClary, *Feminine Endings* (Minneapolis: University of Minnesota Press, 1991) and Simon Frith and Andrew Goodwin, *On Record: Rock, Pop and the Written Word* (New York: Pantheon, 1990).

Robert C. Post
Straightaway Dreams

A few years ago, after I had presented a talk on hot rodding at a German technical museum, the director remarked that research must be difficult because "there is almost no printed literature." Perhaps he meant to say scholarly literature, and that is nearly (though not altogether) true. The quantity of popular periodical literature is overwhelming. One begins with Petersen Publishing's *Hot Rod* magazine, begun in 1948 and still flourishing. Literally dozens of other monthlies have come and gone in that interim, most notably *Drag Racing* (1964–75), a title that was later reincarnated by Petersen (1984–91). The National Hot Rod Association's weekly *National Dragster* is now in its fourth decade, but one should not forget another weekly that was on the scene even earlier, *Drag News* (1955–77). *Bonneville Racing News,* edited by Wendy Jeffries with admirable sensitivity to historical matters, carries on traditions of Veda Orr's annual *Lakes Pictorial* which preceded even *Hot Rod.* A comprehensive collection of periodicals concerned with straightaway speed is held by Don Garlits's Museum of Drag Racing in Ocala, Florida.

The most important scholarly monograph is H.F. Moorhouse, *Driving Ambitions: A Social Analysis of the American Hot Rod Enthusiasm* (Manchester University Press, 1991). Moorhouse, a sociologist, has also published several articles on hot rodding. Other sociological analyses include Gene Balsley, "The Hot Rod Culture," *American Quarterly* 2 (1950): 353–58, and James P. Viken, "The Sport of Drag Racing and the

Search for Satisfaction, Meaning, and Self" (Ph.D. diss., University of Minnesota, 1978). From a historian's perspective, my own articles include "The Cars Won't Fly," *Air and Space* 1 (August–September 1986): 76–84; "Von Daimler und Benz zu Garlits und Beck," *Kultur & Technik* 10 (Heft 2, 1986): 114–23; and "In Praise of Top Fuelers," *Invention and Technology* 1 (1986): 58–63.

Two popular histories worth reading are Wally Parks, *Drag Racing: Yesterday and Today* (New York: Trident Press, 1966), and Dave Wallace, Jr., *Petersen's History of Drag Racing* (Los Angeles: Petersen Publishing Co., 1981), and one should also see two autobiographies: Mickey Thompson with Griffith Borgeson, *Challenger: Mickey Thompson's Own Story of His Life With Speed* (New York: Signet Key, 1964), and *"Big Daddy"—The Autobiography of Don Garlits* (Ocala: Museum of Drag Racing, 1990). Recent books on Bonneville and land speed racing include Peter J.R. Holthusen, *The Land Speed Record* (Newbury Park, Calif.: Haynes Publications, 1986) and George D. Lepp, *Bonneville Salt Flats* (Osceola, Wisc.: Motorbooks International, 1988). In Fallbrook, California, Donald R. Montgomery has published under his own imprint *Hot Rods in the Forties* (1987), *Hot Rods as They Were* (1989), and *Hot Rod Memories* (1991). These are essentially photo albums but not insignificant in their historical analysis.

Enthusiasm is central to the interpretive stance of several major historians of technology, including Brooke Hindle, Thomas Hughes, and Eugene Ferguson. Two of Ferguson's *Technology and Culture* articles provide a good introduction to this literature: "Toward a Discipline of the History of Technology," vol. 15 (1974): 13–30, and "The American-ness of American Technology," vol. 20 (1979): 3–24.

Michael L. Smith
Planetary Engineering

For an overview of some ideas addressed in my forthcoming book "Special Effects," see my article "Selling the Moon," in Richard W. Fox and T.J. Jackson Lears, eds., *The Culture of Consumption: Critical Essays in American History, 1880–1980* (New York: Pantheon, 1983). Much of my research for this essay came from the pages of *Popular Mechanics,* and a number of other mass-circulation magazines of the postwar decades. I have tried to read the messages about technology and society in these magazines alongside the ongoing history of the period—particularly the social role of nuclear and space technologies.

For discussion of the cultural and political attributes assigned to nuclear technology, see Paul Boyer, *By the Bomb's Early Light: American Thought and Culture at the Dawn of the Atomic Age* (New York: Pantheon, 1985); Spencer Weart, *Nuclear Fear: A History of Images* (Cambridge, Mass.: Harvard University Press, 1988); and Michael L. Smith, "Advertising the Atom," in Michael J. Lacey, *Government and Environmental Politics: Essays on Historical Developments Since World War Two* (Washington, D.C.: Wilson Center Press, 1989).

For the history of the space program, see Walter A. McDougall, . . . *the Heavens and the Earth: A Political History of the Space Age* (New York: Basic Books, 1985). Background material on the 1964–65 New York World's Fair can be found in *Remembering the Future: The New York World's Fair from 1939 to 1964* (New York: Rizzoli International Publications, 1989).

Readers who wish to explore connections between political and cultural developments during the postwar period may wish to consult Elaine Tyler May, *Homeward Bound: American Families in the Cold War Era* (New York: Basic Books, 1988); and Thomas Hine, *Populuxe* (New York: Alfred A. Knopf, 1986). For an unusual but highly original study of the era, see Dale Carter, *The Final Frontier: The Rise and Fall of the American Rocket State* (London: Verso, 1988).